D0918994

Knowledge and Power

Knowledge and Power

TOWARD A POLITICAL PHILOSOPHY OF SCIENCE

Joseph Rouse

Cornell University Press

Ithaca and London

First published 1987 by Cornell University Press.

International Standard Book Number 0-8014-1959-x
Library of Congress Catalog Card Number 87-47604

Printed in the United States of America

Librarians: Library of Congress cataloging information appears on the last page of the book.

The paper in this book is acid-free and meets the guidelines for permanence and durability of the Committee on Production Guidelines for Book Longevity of the Council on Library Resources.

Contents

Preface

UNDOUBTEDLY, many of the most profound differences between the world we inhabit and the world of only two centuries ago can be traced back to the intellectual and practical achievements of the natural sciences. Within the setting of the university, attention is perhaps most immediately drawn to the intellectual changes wrought by the sciences. It is not only that we now know so much more about the natural world. Many of our most central *fields* of knowledge did not exist in recognizable form in the mid-eighteenth century: electromagnetism, thermodynamics, evolution, genetics, historical geology, virtually the whole of modern chemistry, and anything whatever to do with the structure and behavior of molecules, atoms, and nuclei. But the physical and social changes brought about by these developments have been even more striking. The face of the globe has been physically transformed by the deployment of technical innovations traceable to developments in the natural sciences. Our most ordinary patterns of life and social interaction, our hopes and our fears, the things we deal with, the problems we confront and the goals we aim for have been still more profoundly affected. We live in a world that has been substantially reconstructed because of what has happened in a very short time within the laboratories of science.

Oddly, a casual perusal of the considerable professional literature in the philosophy of science suggests that little philosophical attention has been devoted to these physical and social effects of the growth of the sciences. Such a casual conclusion is of course misleading in an important respect: the dramatic effects of the growth of the sciences help account for the central place accorded the philosophy of science among contemporary philosophical topics. And if the broader cultural significance of the sciences is given little thematic attention, this significance has perhaps been overlooked only because it is taken as an understandable outgrowth of the overwhelming success of the sciences in their own terms. The *epistemic* achievements of the sciences have been thought to account for the subsequent applications of their theories and techniques. What has seemed to call for sustained philosophical attention has been what this epistemic success has consisted in and how it has been achieved.

In another sense, however, the casual impression is quite correct that philosophers of science have been largely unconcerned with the extension of scientific practices and achievements outside the laboratory. It cannot be taken for granted that increased knowledge of the natural world would inevitably be usable in a significant way or that it would in fact be used. The main currents of philosophical thought about science have not effectively posed the question why the flood of new scientific insights in the past two centuries has been so readily and extensively applicable. Nor have they considered why these possibilities were so enthusiastically taken up within the larger society, or what their social and political impact has been.

Such judgments about the contemporary state of the philosophy of science are complicated by the fact that there have been two quite separate strains of philosophical reflection upon science in the twentieth century.[1] Within the English-speaking world, philosophers have made common cause with the sciences. Scientific investigation in its modern guise has been taken as the model of successful inquiry, and philosophers have been concerned to work out explicitly the epistemic aims, methods, and forms of reasoning it has involved. Philosophers have assumed the task of critical reflection upon the sciences, but their reflections have been situated almost entirely within the scientific project itself. They have been more typically concerned to ask whether some of the more dubious disciplines have lived up to the ideals of the sciences than to raise critical questions about those ideals themselves.

Relations between philosophy and the sciences have been rather different in much of Continental Europe, and philosophers' attitudes toward the sciences have been much more ambivalent. The question whether the methods and results of the natural sciences should be

1. A concise and interesting discussion of the most salient differences between these approaches is provided in Gutting (1979).

more widely emulated or applied has been of pivotal interest, and it has often been answered in the negative. More generally, the concern has been to situate the sciences with respect to other social interests and practices. Political engagement, social and psychological theories, and the arts have been more central to European philosophical concerns than have the natural sciences and mathematics. The sciences have been more at issue for their relation to and influence on these other enterprises than for their own sake. The promise, or the threat, of extending into these other fields the concepts, methods, and patterns of thought characteristic of the natural sciences is the issue that has organized most European philosophical reflection upon science. Even here, however, it is primarily the intellectual and epistemic influence of the sciences that has been most at issue.

Each of these traditions has found reasons not to take the other seriously. From a European perspective, Anglo-American philosophy of science has seemed uncritical and positivistic. It has neglected the fundamental issue of the place of science in any larger social context and has focused inordinate attention on narrow issues of little interest outside the culture of science. The allegedly excessive formalism of its methods has been thought to reflect its lack of critical substance. From across the Channel and the Atlantic, however, the European writers have too often seemed naive and ill informed about the sciences, obscurantist in their prose, and shockingly sloppy in their arguments. Their supposedly critical stance toward the scientific project has often been taken instead to amount to an unreflective hostility toward a more rigorous and progressive mode of thought that they find threatening. Some of their more prominent theoretical perspectives, most notably Freudian psychoanalysis and Marxism, have likewise been discounted as unscientific and objectionably ideological.

These dismissals on both sides are simplistic and profoundly mistaken. Both traditions have important contributions to make to our understanding of the sciences, and they interact in informative ways. They cannot be treated simply as engaged in different projects with little to say to one another. This book reflects my commitment to breaking down some of the barriers that have insulated philosophers of science from one another and obscured our understanding of the sciences.

The overall project of the book is criticism in the European sense. I want to understand the sciences not just as self-subsistent intellectual activities but as powerful forces shaping us and our world. We cannot justifiably believe that the growth of the sciences has simply added more knowledge of natural processes to an otherwise unchanged world of human relationships. Nor can we simply assume that the social and cultural effects of this unprecedented growth have been uniformly progressive. We must ask how the natural sciences have changed us and in what terms we can best critically understand and assess these changes.

At the same time, I have been drawn to the philosophy of science out of a deep respect for, and fascination with, the achievements of the natural sciences. I do not believe we can acquire a more adequate grasp of the place of those achievements in a larger cultural context without a better understanding of how scientific research is done, what it aims for, how it achieves its aims, and how its achievements are recognized. Many of the most interesting critical studies of the sciences have been marred by naive misconceptions of how science is done and what it accomplishes. To rectify these misconceptions, we must take account of what has been learned in the predominantly Anglo-American attempts to explicate the scientific project from within, and we must make further contributions to those attempts.

My desire to overcome the barriers between different philosophical approaches will be evident throughout the book. I have not been concerned with exposition and commentary on the work of other philosophers except when it seems to advance the understanding of the sciences and their social and political context. Nevertheless, I have drawn freely upon a wide variety of sources for contributions to my arguments. The entire project has been most deeply influenced by the works of Heidegger and Foucault, but my reading of them has been aided considerably by study of Quine and Davidson. I undoubtedly would not have seen the same implications in their work for understanding the sciences without the guidance of Kuhn, Hacking, Cartwright, and Hesse or the context provided by contemporary Anglo-American debates over scientific realism.

I found, however, that I could not simply bring together the best of Anglo-American and Continental approaches to understanding the sciences. There are also dangerous tendencies that run through much of the philosophy of science on both sides of the English Channel. Philosophers have too often focused on the narrowly intellectual aspects of science—on scientific theory, the patterns of thought it requires, the kinds of evidence that should incline us to believe it, and the intellectual satisfactions it offers. In this context it is easy to forget that scientific research is also a deeply practical activity. By this I mean not that it is oriented toward applications, but that practical skills and manipulations are crucial to what it accomplishes in its own terms. This is obviously true for experimentalists, whose practical achievements provide most of the material for theoretical work; yet these technical skills and exploits are rarely given their due philosophically.[2] Even theory is more practical and craftlike than philosophers usually recognize. Thus, the problem is not the neglect of one part of science (experiment) in favor of another (theory), but rather a distorting perspective on the scientific enterprise as a whole. The same problem is apparent in most critical studies of the sciences, which tend to look more to scientific influences on other ways of thinking than to the material effects of the sciences on our everyday activities and social interactions. I have found that this approach overlooks some of the most important issues posed for us by the modern developments in the sciences. A central theme in this book is therefore the importance of regarding the sciences, both epistemically and politically, as a field of practical skills and activities and not just as a field of beliefs and reasons.

In this respect I have learned a great deal from recent work in the sociology of science, which meshes in interesting ways with what I have taken from Heidegger and Foucault and from Kuhn, Hacking, and Cartwright. It was not so long ago that the "discovery" of the newly professionalized discipline of history of science caused profound upheavals within mainstream philosophy of science. The sociologists may yet create comparable, and comparably productive, turmoil. But here again my interests may be unusual. Most philosophers who read the sociological literature seem to cite the Edinburgh school of Bloor, Barnes, and Edge—usually unfavorably. I have found the work of Latour, Woolgar, Knorr-Cetina, Pickering, and Pinch more illuminating. I have also found the growing literature on feminist interpretations of science important and provocative, but it is rarely given adequate treatment by other philosophers.[3]

My argument itself contains two central themes. The first is an emphasis, already mentioned, on science as a field of practical activity. The second theme is that we cannot readily separate the epistemological and political dimensions of the sciences: the very practices that account for the growth of scientific knowledge must also be understood in political terms as power relations that traverse the sciences themselves and that have a powerful impact on our other practices and institutions and ultimately upon our understanding of ourselves.

2. A striking exception to the general neglect of this topic is Hacking (1983).

3. A good sample of recent feminist work in the philosophy of science might include Harding and Hintikka (1983), Keller (1985), Harding (1986), and Bleier (1986).

These two themes are closely intertwined, so that the argument of the book cannot simply be presented sequentially. The opening chapter introduces the question of the relations between scientific knowledge and political power, through a reflection upon the recent history of Anglo-American philosophy of science. There is a view about the relation between power and knowledge that is implicit in most of the Anglo-American studies, but I suggest that some recent developments may have made this view untenable.

To elucidate this point, I set aside the explicitly political issues for a while. The first step is to reconsider the work of Thomas Kuhn, which I do in the second chapter. Kuhn is the most prominent among a small group of writers who in the 1950s and 1960s initiated some fundamental changes in the philosophy of science. I nevertheless argue that the most revolutionary aspects of Kuhn's work, resulting from his emphasis on the practical dimension of scientific research, have been almost universally overlooked by philosophical readers. Kuhn's recovery of the practical dimension gives his work a very new look and prepares the way for a more systematic treatment of science as a practical activity.

The third chapter provides a theoretical framework for my subsequent interpretation of scientific practice. The setting is provided by the Diltheyan project of treating the human sciences as hermeneutical rather than empiricist. It has been widely noted that the eclipse of logical empiricism within the philosophy of science has undermined the Diltheyan distinction by suggesting that all knowledge is hermeneutical, even in the natural sciences. What has not commonly been noticed is that such a universal hermeneutics can take two forms, depending on whether interpretation is likened to translating a sentence or to engaging in a practice. This distinction between a Quinean theoretical hermeneutics and a Heideggerian practical hermeneutics[4] turns out to parallel that noted in chapter 2 between the two readings of Kuhn. The differences are subtle but important. They shape the rest of the book, because I adopt the standpoint of a practical hermeneutics in my subsequent interpretations of scientific research.

The core of my interpretation of the sciences as fields of interpretive practice is developed in chapter 4. I begin with a criticism of Heidegger's early philosophy of science, because there he mistakenly tries to exempt the sciences, at least partially, from his practical hermeneutics. I argue that such an exemption requires us to

4. Dreyfus (1980) draws essentially the same distinction, calling the two possibilities "theoretical holism" and "practical holism."

overlook both the engaged, opportunistic character of scientific research and also the concrete, contextual nature of scientific theorizing. This oversight causes us to misunderstand the role of experiment in the sciences and, with it, the significance of the local, material setting of the laboratory, clinic, or field site and the technical, practical know-how acquired within such settings. These misunderstandings are made possible by a misinterpretation of the ways the materials and practices of research are standardized for translation into new local settings. They are also reinforced by a disregard for the social setting within which the significance and epistemic standing of scientific claims are adjudicated. When we take these considerations into account, we find in science a local, existential knowledge based on a practical grasp of a configuration of equipment, techniques, social roles, and intelligible possibilities for employing them.

Many philosophers of science today are likely to regard the account in chapter 4 as unacceptably anti-realist in its implications. Realism is the view that scientific theories are true or false depending upon whether the objects they refer to actually exist and have the characteristics the theories ascribe to them. It has become the predominant view within Anglo-American philosophy of science—largely, I believe, because of the recognized deficiencies of the anti-realist alternatives. In chapter 5 I argue that both realism and the most prominent forms of anti-realism are unacceptable. They share unarticulated views about the relations between meaning, existence, and truth that do not stand up to criticism. When these views are recognized and rejected, we are able to accept the commonsense notion that science tells us what the world is like without our having to accept either the metaphysical extravagances of scientific realism or the implicit use of realism as a justification for exempting the sciences from political criticism.

Readers unfamiliar with recent philosophical discussions about realism may find chapter 5 hard going. If the question whether scientific theories tell us what the world is *really* like has never particularly concerned you, you can probably skip this chapter without losing the thread of the argument. Indeed, if my argument in the chapter is sound, you were right not to be bothered about this issue in the first place.

Chapter 6 returns us to the political themes of the book. It does so initially by returning to the Diltheyan distinction between the natural and the human sciences. There have been several recent attempts to resurrect Dilthey's distinction in a new guise, including arguments by Hubert Dreyfus and Charles Taylor that rely upon the kind of practical hermeneutics that has been central to my claims in this book. These neo-Diltheyan projects are important because they aim to show that in the human sciences there is something at stake politically and culturally that is not at issue in the natural sciences. If this were true, the natural sciences would not be as important to us politically, but they would also be exempt from certain kinds of political criticism. This chapter shows why the various neo-Diltheyan arguments are untenable and why we must regard the natural sciences as fundamental to our self-understanding and to what is at issue politically in our lives.

It is not enough just to claim that the sciences are politically important, however. We must show what kinds of political effects they have and in what terms these effects must be understood. In chapter 7 I argue that the concept of "power" must be fundamental to a political interpretation of the sciences. I have already claimed in the first chapter that the sciences compel us to rethink our traditional interpretations of this concept. Foucault's analysis of the forms of "disciplinary power/knowledge" that became prominent in the eighteenth century turns out to be quite useful in this respect. Scientific laboratories offer analogues to many of the power relations that he discovers in prisons, hospitals, barracks, factories, churches, and schools. The power relations that traverse scientific practices themselves are also extended outside the laboratory. The construction of simplified and controlled "microworlds" within laboratories provides models and strategies for reconstructing the world around us. This reconstruction changes the political possibilities open to us and creates new issues we have to respond to. Understood in these terms, the growth of the natural sciences presents us with some of the most important political issues we face today, issues that cannot be separated from the epistemological concerns that have traditionally been the focus of the philosophy of science.

A political philosophy of science should provide a framework within which the political impact of scientific practices and their extension outside of the sciences proper can be critically assessed. Providing such a framework is beyond the scope of my argument here; it would require a much more substantial political context than I have been able to provide. Nevertheless, the arguments in the book point us in the direction of a substantive political critique of the sciences. In the Epilogue, I suggest some alternative possible directions for such a political assessment and sketch out some of the issues at stake in pursuing them. These issues both impel philosophy of science in new directions and provide an important new context within which to consider the traditional issues in the field. If we are to understand what is at stake in the relentless pursuit of more scientific knowledge, we must consider scientific practices themselves more intensively and also look outward at their effect on the world around us.

A project like this one could not easily be completed without the help of many people and without considerable institutional support. I have incurred more debts in the course of writing this book than can be explicitly acknowledged here, but I welcome the opportunity to express my gratitude for some of the most prominent contributions.

I wrote a large part of the manuscript with the support of a grant-in-aid from the American Council of Learned Societies and a sabbatical from Wesleyan University, which enabled me to spend a most productive and enjoyable year in residence at the Center for the Study of Values at the University of Delaware. Preparatory work for writing the book was supported by a Summer Seminar Fellowship from the National Endowment for the Humanities, which allowed me to participate in Hubert Dreyfus's seminar on Heidegger and Foucault in 1983; a Summer Faculty Research Grant from the University of Maine at Orono; and a Summer Fellowship from the National Endowment for the Humanities. I am most grateful to the Council, the Endowment, the University of Maine, and Wesleyan for their generosity and for their encouragement during the long and difficult period when my ideas were taking shape. The hospitality offered by the Center for the Study of Values, its director Norman Bowie, and its staff of Patricia Orendorf and Sandra Manno provided me with a congenial and supportive institutional home during the trying times of first committing this work to paper.

Many friends and colleagues have stimulated and challenged me and have contributed ideas that eventually found their way into the book. I could not possibly mention them all, but I am especially indebted to Hubert Dreyfus and Mark Okrent. My conversations with them, including our many disagreements, and my reading of their work have left their mark upon much of what I have to say here. This will no doubt be apparent to them at various points, but their influence runs throughout the book. I also appreciate the contributions of Douglas Allen, Patrick Heelan, Karin Knorr, and Samuel Todes, some of whom may not realize just how influential they have been.

My thinking about how science is done has been immeasurably sharpened and clarified by conversations over the past thirteen years with many friends and colleagues in the sciences. Their patience in explaining their work and in answering my questions has been much valued.

Many of the ideas I develop here were first worked out in my classes at Wesleyan University and the University of Maine at Orono. To the students who endured my first halting steps and missteps toward this work, and whose questions, comments, and papers helped focus what I have to say, I am grateful.

Earlier versions of chapter 1 were presented at Dickinson College, Williams College, and Bates College. I am much obliged to the audiences at these presentations for their sympathetic response and their helpful questions and objections.

I have especially benefited from many critical readings of individual chapters of the manuscript. I offer special thanks to Brian Fay, Mark Okrent, Sue Fisher, Howard Bernstein, Jana Sawicki, Christopher Gauker, and the two anonymous reviewers for Cornell University Press, whose comments and criticisms have made this a much better book.

I have benefited for almost three years now from the attention and encouragement of a wonderful editor, John G. Ackerman of Cornell University Press. It has been a pleasure to work with him. Alice M. Bennett's thorough editing of the manuscript caught a number of errors and made the text considerably more readable.

Mary Kelly, Mildred Pugh, and Sandra Manno did much of the work of manuscript preparation. Without their efficiency and cheerful goodwill, the years I have worked on this book would have been much more trying.

No project of this magnitude can succeed without the support and forbearance of family and friends. My parents, J. T. and Helen M. Rouse, have been there to help me in countless ways throughout the long road here. During the difficult time when I was working on the manuscript, Annie V. F. Storr provided me with encouragement, support, and love. To her this book is gratefully dedicated.

JOSEPH ROUSE

Middletown, Connecticut

Knowledge
and Power

Chapter 1

Introduction: Knowledge and Power

DEVELOPMENTS in the philosophy of science over several decades have effected a striking reversal of many of the central tenets of the logical empiricist account of science that once held sway throughout the English-speaking philosophical world. These developments have in turn been linked by some to a more general philosophical attack upon the seventeenth-century tradition in epistemology.[1] At the same time, however, post-empiricist philosophy of science implicitly reveals just how closely traditional interpretations of knowledge were connected to an understanding of power and of the relation between power and knowledge. Although the retreat from empiricist epistemology has been widespread, there has been a much greater reluctance to abandon the associated interpretation of power and its relation to the achievement of knowledge. A central concern expressed in many recent philosophical discussions of science has been to find a way to preserve the received understanding of the relation between power and knowledge without endorsing the empiricist account of knowledge that has often been used to sustain it. Larry Laudan's response to some of the more radical critics of empiricism captures this concern succinctly:

> Both Kuhn and Feyerabend conclude that scientific decision-making is basically a political and propagandistic affair, in which prestige, power, age, and polemic decisively determine the outcome of the struggle be-

1. Rorty 1979; Bernstein 1983; Toulmin 1972.

> tween competing theories and theorists. . . . Having observed, quite correctly, that the Popperian model of rationality will do scant justice to actual science, they precipitately conclude that science must have large irrational elements, without stopping to consider whether some richer and more subtle model of rationality might do the job.[2]

We may have to revise or reject particular models of knowledge or rationality, Laudan thinks, but we cannot do without the underlying distinction between beliefs rationally (or scientifically) arrived at and those materially influenced by the use of power and polemic. Providing a well-founded justification for this distinction seems to many to be the definitive task for a philosophy of science. Richard Rorty suggests why this aspect of the epistemological tradition has been so important: "We are the heirs of three hundred years of rhetoric about the importance of distinguishing sharply between science and religion, science and politics, science and art, science and philosophy, and so on. This rhetoric has formed the culture of Europe. It made us what we are today."[3] Our understanding of science and its place in our culture is far more deeply influenced by its presumed opposition to "the two-fold influence of tyranny and superstition"[4] than by any particular epistemology.

What is not generally realized, however, is that this opposition between the achievement of scientific knowledge and the deployment of power not only has rested upon a dubious epistemology, but has also presupposed a disputable interpretation of what power is and how it figures in our understanding of the world. Perhaps a revision of our understanding of power will change the configuration of knowledge and power more fundamentally than have the recent developments in epistemology and the philosophy of science. Perhaps even our understanding of knowledge and rationality will be decisively affected by what may at first appear to be strictly an issue in political philosophy. To see how this might be the case, we need to consider first what has happened within epistemological philosophy of science. For it is the space of epistemological discussion of science that has been formed in opposition to what is political and has thereby defined the political by its exclusion.

Pragmatism and a New Empiricism

According to the tradition, knowledge consists in accurate representation whose accuracy is recognized and justified by the knower. Science

2. Laudan 1977: 4.
3. Rorty 1979: 330–31.
4. Condorcet 1955: 124.

has been important to philosophers as a perspicuous example of the achievement of knowledge, "commonsense knowledge writ large."[5] Science is a means, the most successful means we have devised to date, for constructing and improving representations of the world. The task of a philosophy of science has been to interpret, explain, and defend that claim to success. In the English-speaking world, the predominant approaches to that task have been empiricist. The success of science has been attributed to the testing of its theoretical representations against carefully controlled observation and reporting of relevant features of the world, followed by the revision or replacement of those theories that fail to correspond to what is observed. Such a representational model of knowledge presupposes the possibility that there is a gap between the way the world is and our representations of it. The world is independent of how we represent it, and our representations may depict the world inaccurately. Scientists aim to close this gap by improving our representations; philosophers play a critical role in this project by reflecting upon whether scientists' procedures are adequate to achieve such closure and to recognize it.

Representationalist models of knowledge have always faced the problem of how we are to gain access to the things our representations are supposed to correspond with. Empiricists have responded to this difficulty with the claim that the senses give us access to the way things are, independent of how we represent them. Carefully controlled observation can thus provide the independent check on our representations that allows us to work to improve their accuracy. Without such an independent touchstone, it might never be clear when we were improving the accuracy of our representations and when we were making things worse. We could not even justifiably claim that science was making progress, let alone explain how such progress was made. With such independent access, scientists could reasonably expect to continue to develop more accurate descriptions of how the world is, as long as they were careful in their testing, dutiful in abiding by the results, and imaginative in constructing alternative theories to replace those that failed. The prephilosophical belief that science steadily progresses seems to be justified and reinforced by such empiricist interpretations.

Many of the challenges to empiricist philosophy of science have focused upon the presupposition that we have direct observational access to the way the world "really" is. The criticism has been threefold: (1) When scientists observe, they do so selectively, and these

5. Popper 1953: 22.

selections are governed by their theoretical (and sometimes practical) interests.[6] (2) Scientific observation and theory testing are complicated activities, employing complex apparatus and relying upon a host of auxiliary theories and prior determinations and assumptions of other facts; there is no such thing as the "direct" comparison of a scientific theory with observed facts.[7] (3) Even if we could devise simple tests, we cannot directly compare a scientific theory with what we observe. We must first describe what we observe, so that we can compare one statement with another, but we cannot describe what we observe without making use of theoretical assumptions built into our concepts. Even the simplest concepts, such as "yellow" or "ball," have been said to involve far-reaching theoretical assumptions.[8] Thus scientific observation of the world is theoretically selected and interpreted and functions only within a network of presupposed theories. Observation is very far from giving us an independent check on the accuracy of our theoretical representations.

These challenges have led to a revised picture of how science develops. According to this "post-empiricist" model, scientists compare their theoretical representations with other theoretical representations rather than with the observed, uninterpreted world. The history of science is not a story of the gradual accumulation of a storehouse of knowledge about the given world. It tells instead of discontinuous changes in the overall structure of our representations and, with them, of changes in how the world appears to us. New theories sometimes radically reconceptualize the world, replacing some of our previous knowledge rather than just adding to it.[9] This revised picture of science has had some remarkable successes, both in resolving the many embarrassing conceptual difficulties in empiricist philosophy of science and in developing a fruitful dialogue between historians and philosophers of science. It has also focused attention on two central issues and has raised important questions about two others that have been less often discussed.

The most widely discussed issue results from the recognition that once we abandon the notion of measurable correspondence with the observed world, it is not so clear what counts as a "better" theoretical representation. Recent philosophy of science has been haunted by the fear that the history of science will turn out to be not a story of relentless progress, but merely one damn theory after another, with

6. Kuhn 1970a, chaps. 6, 9; Hanson 1958, chap. 1.
7. Lakatos 1978: 43–46; Kuhn 1970a, chap. 9; Hanson 1958, chap. 1.
8. Hesse 1980: 65–83.
9. The locus classicus for this claim is Kuhn (1970a, chaps. 1, 9–13).

no rational reason to prefer more recent theories to their predecessors. As a result, the philosophical literature is replete with attempts to demonstrate the rationality, progress, or growth of science without having to appeal to increases in the accuracy of theoretical representation as a criterion.[10] None of these attempts has been acknowledged as entirely successful—all the more remarkable because the *need* for a successful account has been rather generally felt.

The second widely discussed issue is the question of realism: Do scientific theories tell us what kinds of objects there really are in the world and how they actually behave, or are theories only devices for calculation and prediction that help us make sense of what we observe without literally describing it? Empiricists had generally been realists about observation statements and anti-realists about theories, but the collapse of any sharp distinction between theoretical and observational statements seems to make this combination of views untenable. Most contemporary philosophers of science have turned toward realist interpretations of theory while usually granting that we do not yet know *which* theories actually correspond to the way the world is, or at least in which respects our best theories most closely approximate a true correspondence. A minority have adopted instead an anti-realist interpretation of *all* scientific discourse, including observation reports, while a few have attempted to preserve something akin to the traditional empiricist distinction. The most telling points in favor of the first two approaches have often seemed to be their criticisms of the alternative, which has made the issue of realism something of an embarrassment to many philosophers of science.

The third issue, less often discussed, concerns what to make of the relations between the natural sciences and the disciplines that study human beings, their artifacts, and their institutions. The latter disciplines are sometimes thought to be united by a concern for interpreting meanings, whether of texts, objects, practices, or institutions. The natural sciences, by contrast, were supposed to study natural processes independent of whatever meanings are ascribed to them. Meaning is allegedly constitutive of social life but extrinsic to the workings of nature. When one claims that even the natural sciences study nature only as theoretically interpreted, however, this distinction seems to collapse. The same kinds of interpretive procedures we use to understand the social life of others also apply to nature as they, and we, encounter it. Witchcraft among the Azande and physics among the Europeans and Americans become comparable; both count as ways

10. See, for example, Lakatos 1978; Laudan 1977; Newton-Smith 1981.

nature manifests itself within a theoretical framework.[11] Of course this conclusion may not present any *problems* at all. Cultural anthropologists have been relativists about cosmology for a long time, and perhaps philosophers should simply join them. What may make it a problem, however, is that there do seem to be significant differences in style, subject matter, rhetoric, and especially results between the natural sciences and the human studies. These differences seem to be readily explainable within traditional epistemology, whether one takes them as a reflection of essential differences between studying human beings and studying nature or simply as a sign of the immaturity of the human studies. Post-empiricists either must decide nevertheless that these often-recognized differences have no significant epistemological basis or counterpart or must find a new basis for them. Otherwise post-empiricist models of natural science themselves seem to need revision.

A final difficulty emerges when we recognize that post-empiricist models of science as sketched above seem to provide no natural account of the connections between science and technological capability. How and why have the growth and development of the sciences contributed to the vast expansion of the technological power at the disposal primarily of the Western nations and their institutions? One might challenge the extent of the connection between science and technology, arguing that technological development has a history largely independent of the sciences. But there are limits to any challenge on these grounds, for science has its own history of increasing technical control of phenomena within the laboratory. Much as we may respect the intellectual achievements of four centuries of modern science, surely its technical achievements are as much or more in need of explanation. Modern science and technology have, I think, transformed the globe more thoroughly than they have transformed our thinking about it; an eighteenth-century intellectual would be far more at home in our university classrooms than in our homes, factories, and cities. To discuss modern science without centrally accounting for its technical capabilities must seem, if not bizarre, at least a little odd. Yet as Hilary Putnam once put it, non-realist accounts of science (such as the post-empiricist model described above) seem at first glance to make the technical success of science a miracle.[12] We want to know why some ways of talking about the world assist us in manipulating it effectively and others do not; standard post-empiricist accounts have not answered this question very satisfactorily.

11. Cf. Wilson 1970; Hollis and Lukes 1982.
12. Putnam 1978: 19.

There have been three important strategies for responding to these difficulties. One, commonly called "convergent realism," has focused almost entirely upon the second of these four issues, and consideration of it can be safely postponed to chapter 5, when we take up in detail the questions concerning realism. Of the remaining two, the first is rooted in American pragmatism.[13] The pragmatists appeal to the outcome of critical discussion by a community of inquirers as the arbiter of questions about truth and reality. Truth is what would emerge as the result of unconstrained inquiry pursued indefinitely. Even our methods and standards of evaluation cannot be established in advance, for they too can be improved in the course of inquiry. Inquiry, of course, never actually comes to an end, and we can never be certain that any particular result will not be overturned by further investigation. But we need entertain no serious fears that our inquiries may fail to disclose things as they really are: the way things show themselves to our ongoing inquiries simply *is* the way they really are. Richard Bernstein has nicely summarized this approach to understanding scientific rationality, saying that "central to this new understanding is a dialogical model of rationality that stresses the practical, *communal* character of this rationality in which there is choice, deliberation, interpretation, judicious weighing and application of 'universal criteria,' and even rational disagreement about which criteria are relevant and most important."[14]

This understanding of the openness of theoretical discourse enables pragmatists to dismiss disputes about realism, and about metatheoretical criteria for the adequacy of theories, as pseudoproblems. To take these problems seriously is to worry that all our theories might be false, all our terms fail to refer, or all our conversation miss the point. But to do so, they claim, is to fail to grasp that truth and falsity, reference and failure, cogency and pointlessness make sense only within the horizons opened by the actual development of inquiry—the "conversation," as Richard Rorty has called it. The worry over the possibility of total failure can be dismissed as incoherent or vacuous unless it can be connected to the specific inadequacies of specific theories or disputes.[15] But such specific criticisms bring us

13. I have in mind such philosophers as Rorty (1979, 1982), Bernstein (1983), Putnam (1981), Goodman (1978), and Habermas (1968a, 1971), who all claim links to the tradition of American pragmatism; some of them would obviously dispute some or all of the points I formulate as "pragmatist," which are closest to Rorty's and Bernstein's positions. This is the hazard of indicating general trends, but I think the similarities in approach are still there and are significant.

14. Bernstein 1983: 172.

15. Bernstein 1983: 223–31.

back into the conversation, require us to take specific stands, and thereby remove the general worry with which we began. The skeptical worry is said to take us out of the serious conversation within which theories and standards actually conflict. Getting back in dissolves the skeptical worry.

Such a pragmatism also allows us to dismiss demands for an explanation of the differences between the natural sciences and the human studies and for an account of the technological success of the natural sciences. It may well be that our current theories and argument patterns in the natural sciences have important similarities to one another and that they differ in these same respects from the disciplines that study human beings. But there can be no nontrivial explanation of these differences that does not make them just a peculiarity of our culture. To say that their objects of inquiry are different in essential respects is only to appeal to the theories whose differences we are supposed to explain. To say that the theories themselves or the methods by which we arrive at them differ is only to restate the problem. The assertion of a difference may be important to us, but it can be no more than a feature of our current interpretations, not a general criterion for any acceptable interpretation. For it might be just with respect to its apparent differences from the human studies that the subsequent development of the natural sciences will depart from ours. Richard Rorty has made this claim forcefully:

> But for all we know, it may be that human creativity has dried up, and that in the future it will be the *non*human world which squirms out of our conceptual net. It might be the case that all future human societies will be . . . humdrum variations on our own. But contemporary science (which already seems so hopeless for explaining acupuncture, the migration of butterflies, and so on) may soon come to seem as badly off as Aristotle's hylomorphism. . . . The line . . . between that portion of the field of inquiry where we feel rather uncertain that we have the right vocabulary at hand and that portion where we feel rather certain we do *does,* at the moment, roughly coincide with the distinction between the fields of the *Geistes–* and the *Naturwissenschaften.* But this coincidence may be *mere* coincidence.[16]

Similar issues arise when we consider why the natural sciences have contributed to technological control in such spectacular ways. The relative power of physical and biological technologies may of course conceivably be explained in one way by social or political factors: we

16. Rorty 1979: 351–52.

are interested in controlling physical processes, or we permit it (in ways that would be unacceptable for social technologies), or we have chosen for description those physical situations most amenable to control. Similar accounts might also be given to explain why the human studies and the natural sciences have developed differently in our culture. But once again we cannot explain, without invoking the same theories whose technological implications are supposedly in need of explanation, why just these features of the world are so amenable to representation and control. The success of these procedures remains a brute fact.

We have seen that pragmatism is not without resources for responding to the principal issues I cited as facing a post-empiricist philosophy of science. Nevertheless, many philosophers of science have not been satisfied with such responses, and this dissatisfaction has motivated attempts to develop a more adequate empiricist understanding of science. Under the rubric "new empiricist," I group such diverse philosophers as Ian Hacking, Nancy Cartwright, Mary Hesse, and in some respects Larry Laudan.[17] They differ on many important issues, but there are several common themes in their work. These new empiricists do not attempt to resurrect the distinction between theoretical and observational statements or languages, nor do they generally defend a naive realism about observation. The world as we encounter it is already interpreted, and theories play an important role in our received interpretations. Furthermore, they recognize that there have been significant discontinuities in the historical development of scientific theories. Science cannot claim to provide an accumulating store of theoretical knowledge. Nevertheless, they argue in various ways, science has exhibited empirical progress across its theoretical revolutions, even though the interpretations of what that progress consists in have changed. Hesse succinctly summarizes this central insight: "Successive theories supersede and reinterpret their predecessors, but without rejecting the empirical discoveries that they embody. . . . Lawlike structures and similarities of nature between physical systems have been maintained and are cumulative. Theoretical interpretations of what the natures of these systems absolutely are, are not."[18] The idea is that we can agree in identifying an empirical phenomenon despite even radical differences in our theoretical interpretations of it. We can also determine whether a theory adequately accounts for such an identified

17. Hesse 1980; Hacking 1983; Cartwright 1983; Laudan 1977. Laudan differs notably from the others by placing equal emphasis upon conceptual improvement and empirical growth, and by being a nonrealist about theoretical entities.

18. Hesse 1980: 176–77.

phenomenon in the theory's own terms without necessarily accepting the theory or even acknowledging the legitimacy of its success. Laudan puts this idea in terms of the problem-solving capacities of various theories: "In principle, we can determine whether a given theory does or does not solve a particular problem. In principle, we can determine whether our theories now solve more important problems than they did a generation or a century ago."[19] When we actually attempt such determinations, the new empiricists tend to point out, it is evident that the empirical capabilities of the sciences have grown dramatically over time: "One can see by direct inspection that knowledge has grown. . . . The point is not that there is knowledge, but that there is growth; we know more about atomic weights than we once did, even if future times plunge us into quite new, expanded, reconceptualizations of those domains."[20]

The new empiricists argue that, all things being equal, what makes for a better scientific theory is empirical growth (or expansion of problem-solving ability). This claim does not involve a comparison of the truth content of the theories, since theories are not to be interpreted (or at least not assessed) realistically. Nor is it a comparison of the accepted empirical phenomena they entail, since theories can adequately cope with phenomena whose description they do not entail and can fail to deal with phenomena whose description they can be engineered to entail.[21] In Cartwright's provocative formulation of the former point, "the fundamental laws of physics lie."[22] Nevertheless, theories replace others because scientists judge them to account for more empirical phenomena than their predecessors did, where such accounting is a matter of interpretation or judgment. We account for phenomena in a variety of ways, as Cartwright, Hacking, and (in a different way) Laudan especially take pains to point out.[23] We accept those theories that seem best to account for them in the ways most important to us. This approach may or may not justify the rationality of science compared with other activities, but it does account for the sense in which scientific theories can be said to represent an improvement over their predecessors.

Except for Hesse and Hacking, the new empiricists have not seriously discussed the relations between the natural sciences and the human studies, but they could provide a fairly straightforward (albeit

19. Laudan 1977: 127.
20. Hacking 1983: 119–20.
21. Laudan 1977: 22–24.
22. Cartwright 1983, essays 2, 3, 6.
23. Hacking 1983: 210–19; Cartwright 1983: 44–53, 143–62; Laudan 1977: 22–25.

not explanatory) account of the differences. With some exceptions in both cases, the natural sciences take the pragmatic criterion ("the adoption of one overriding value, namely the criterion of increasingly successful prediction and control")[24] as their predominant concern, and they generally succeed, whereas the human studies generally fail on pragmatic grounds and often adopt alternative values. Obviously the human studies are not without their predictive successes, but these areas do not exhibit the spectacular growth we find in the natural sciences. Furthermore, the changes that natural scientific knowledge enables us to introduce into the world are enduring, while what we can do with social scientific understanding is often relatively ephemeral.[25]

Although Hesse is sensitive to the dangers of overstating differences between the disciplines, she also offers some tentative possibilities for explaining what differences do seem to exist concerning the application of the pragmatic criterion. Her claim is that many of the human studies will not meet the conditions she finds necessary for successful empirical learning: "There must be sufficient possibility of detailed *test* to reinforce correct learning; the environment must be sufficiently stable for the self-corrective learning process to *converge;* and there must not be such strong action by the [learning process] on its environment that either it exhibits no convergence, or what it learns is just an artifact of the [process] itself."[26] Such conditions will not sharply distinguish the natural sciences from other disciplines, and they certainly provide no explanation of why the line between learnable and unlearnable environments falls where it does. But they do provide a suggestive epistemological distinction between the natural sciences, where these conditions generally seem to hold, and the other disciplines, where one or more are often violated.

Finally, the new empiricism internalizes the connection between the development of science and the improvement of technical capability. The achievement of technical control over natural processes is seen not as a by-product of theoretical development but as its raison d'être. Habermas, who has clearly influenced Hesse on this point, concludes that "the empirical sciences have developed since Galileo's time within a methodological frame of reference that reflects the transcendental viewpoint of possible technical control. Hence the modern sciences produce knowledge which through its *form* is technically exploitable

24. Hesse 1980: 188.
25. Hacking 1984.
26. Hesse 1980: 182.

knowledge, although the possible applications generally are realized afterwards."[27] None of the new empiricists has dealt extensively with what this form (or other characteristic) of scientific knowledge is that opens the possibility of technical control. They are all dissatisfied with the hypothetico-deductive model of explanation as an account of how science leads to successful prediction. Hacking insists that "experimentation has a life of its own,"[28] independent of the development of theoretical hypotheses to guide or explain it. Laudan, we have seen, denies that problem solving can always be reduced to entailment relations.[29] Cartwright argues that the use of bridge principles in hypothetico-deductive explanatory patterns is very limited, that very few bridge principles are actually employed in science, and that these apply only to highly fictionalized situations.[30] Most real cases of explaining or predicting particular phenomena, in physics at least, "abound with problems for the [hypothetico-deductive] account."[31] But having said this, they do not so clearly indicate what will replace the role of fundamental theories in accounting not just for particular events, but also for the success of science in achieving technical control of those events. Cartwright and Hacking both discuss the importance of *causal* laws for this purpose; Cartwright emphasizes the fruitful use of approximations that connect theory to real situations, and Hacking suggestively discusses the relatively autonomous development of experimental practices.[32] But if the new empiricism is to succeed as a view of science, much more attention must be paid to the practices that bring about the empirical growth and technical control that supposedly constitute the growth of scientific knowledge. New empiricist philosophy of science depends on "a Back-to-Bacon movement, in which we attend more seriously to experimental science"[33] and also to how theories are actually employed in acquiring and refining our technical capabilities.

The Received View of Knowledge and Power

We are now prepared to ask how power is understood within these various epistemological interpretations of science and therefore what

27. Habermas 1968b: 72–73, 1970: 99.
28. Hacking 1983: 150.
29. Laudan 1977: 22–24.
30. Cartwright 1983: 144–62.
31. Cartwright 1983: 106.
32. Cartwright 1983: 21–43, 74–86, 128–42; Hacking 1983: 149–66, 220–61.
33. Hacking 1983: 150.

relationship is supposed to exist between scientific knowledge and power. Two, or perhaps three, types of interaction are usually thought to occur between knowledge and power. First, knowledge can be *applied* in order to achieve power. Perhaps more accurately, one might say that certain kinds of power can be more readily achieved or more effectively deployed if one can accurately represent one's situation and the effects of one's instruments. Knowing how things are or how they function creates opportunities for manipulation and control; not knowing can misdirect or thwart attempts to intervene. A second type of relation occurs when power is used to impede or distort the acquisition of knowledge. False beliefs can be imposed or given unjustified plausibility, and true beliefs can be suppressed. Sometimes a third relation is distinguished, that knowledge can liberate us from the repressive effects of power. It can uncover the distortions power imposes and unmask the disguises that permit power to operate with reduced interference. Liberation might be reducible to another kind of "application" of knowledge; it might also be seen as simply the other side of the distortion of knowledge by power; or it might be regarded as a distinctive relation not reducible to the other two.

Power and knowledge remain extrinsic to one another in all these types of interaction. Knowledge acquires its epistemological status independent of the operations of power. Power can influence what de facto is known, but its being known, and what it is for it to be known, cannot be subject to the influence of power. That is, power can influence what we believe, but considerations of power are entirely irrelevant to which of our beliefs are true, which of these are known to be true, and what justifies their status as knowledge. It is generally believed that knowledge is best achieved within an inquiry freed from political pressure, but that ultimately an epistemological assessment of that achievement must not refer to the intervention of power either in support of or in opposition to knowledge. Similarly, power may or may not serve knowledge or draw upon it; it remains power all the same. In their constitution as power and as knowledge, power and knowledge are (in principle) free from one another's influence.

We see this independence most clearly in the asymmetry attributed to the effect of power on the achievement of knowledge. Knowledge can be suppressed or distorted by the use of power, but power cannot contribute constructively to the achievement of knowledge. If we wish to explain which beliefs we or others hold, on the traditional view a reference to the effects of power must be "confined to the pathology of belief: to irrationality, or error, or deviance from rational norms. . . . *Correct* use of reason, and true grounded belief, need no causal explanation, whereas error does need it."[34] No explanation is necessary because the epistemological justification of correct belief is sufficient to explain our acceptance of it and normally takes priority over any other explanation. Reference to the influence of power would be redundant. The realm of epistemological justification is a space of conflicting reasons set off from the struggles of competing powers. This same exclusion occurs with respect to the possible application of knowledge. Before knowledge can be applied it must be achieved, and its achievement is conceptually independent of its subsequent application. What is true (and thus what can be known) does not depend upon what actions we might be able to take once we recognize its truth. Indeed, to fail to recognize this independence is to allow the possible application of knowledge itself to play a distorting role in the evaluation. Consider the way Laudan excludes the possibilities for technological application from any role in the cognitive assessment of knowledge claims. He contrasts the "*cognitively rational weighting* of scientific problems" to

> occasions when a problem becomes of major importance to a community of scientists on nonrational or irrational grounds. Thus, certain problems may assume a high importance because the National Science Foundation will pay scientists to work on them or, as in the case of cancer research, because there are moral, social, and financial pressures which can "promote" such problems to a higher place than they perhaps cognitively deserve.[35]

A similar point can of course be made about the possible liberating effects of knowledge. The truth may set us free or it may not, but it remains truth all the same. The point in each case is the same: power can influence our motivation to achieve knowledge and can deflect us from such achievement, but it can play no constructive role in determining what knowledge is.

This account of the exclusion of power from knowledge has traditionally presupposed something like a representational account of knowledge, but it also contains a specific understanding of what power is and how it works. This view is by no means unique to epistemology or the philosophy of science; its most important features can be found in both the liberal and the Marxist traditions in political philosophy. There are three features I want to draw attention to: first, that power is possessed and exercised by specific agents (per-

34. Hesse 1980: 32. Hesse here summarizes not her own view, but a prevalent one that she goes on to criticize.

35. Laudan 1977: 32.

sons, institutions, classes, groups), usually in centralized positions (the state, corporations, ruling classes) from which power is used centrifugally; second, that power operates on our representations, but not on the world represented; and third, that power is primarily repressive, secondarily enabling, but not productive. Each of these features must be explained and defended.

As an interpretation of traditional beliefs about power, the first point should not be controversial. There may be disagreement about who actually possesses and exercises power in any particular society, whether that exercise of power is legitimate, and whose interests it serves, but not about the basic conception underlying these disagreements. In the contractarian tradition power belongs to individuals and is ceded to the state; Marxists respond that power is alienated from its possessors and turned against them. The concern is ever the legitimacy of power and the worth of the ends or interests it serves.

The second point seems less obvious at first, but it becomes less controversial once it is understood. Consider the following remark by Hesse, which does not seem to represent a controversial claim: "While it may be true that the most powerful group can to a greater or lesser extent impose its will upon the development of the social system, it does not at all follow that the theory informed by its value standpoint gives the true dynamical laws of that system on a pragmatic criterion, or the best theory on any other criterion except that truth resides in the barrel of a gun."[36] The way the world is, is not produced or changed by the exercise of power; only our beliefs about it may be changed. This statement reiterates the point that power can impose belief but cannot produce or guarantee the truth of that belief. In the passage cited Hesse takes on the more difficult case of social knowledge; in the case of knowledge of nature, it seems even more initially obvious that the exercise of power cannot change how the world is. Physics is impervious to the demands of the powerful, however effective these demands may be against physicists. Furthermore, power may operate most fundamentally upon belief through ideological distortion, because applying power directly against persons is notoriously ineffective in producing belief. The threat of death, torture, imprisonment, or other penalty can change overt behavior and thus can compel profession of belief, but this result is a far cry from compelling belief. The apocryphal story of Galileo at his trial makes this point tellingly: having renounced the teaching of Copernicus on pain of execution for heresy, he still mutters to himself, "And yet, it

36. Hesse 1980: 197.

moves." Power may thus seem much more dangerous when it changes beliefs by persuasion or distortion.[37] Thus a central concern of political philosophy has become understanding ideology and contrasting it to beliefs undistorted by this form of power. The same shift is evident among philosophers of science when they consider the effects of power upon science. Recall the passage I cited from Laudan, which challenged the view that science is "a political and propagandistic affair" whose results are determined by "prestige, power, age, and polemic."[38] It is the associations here that are telling. Propaganda and politics, polemic and power: it is in the medium of our beliefs that power is supposed to operate, as polemic, propaganda, ideology. This is not to say that power is never exercised to control what we do rather than what we believe, but it is through beliefs that power primarily impinges upon knowledge. In Mill's classic study of the legitimation of power, *On Liberty,* it is the first section, concerning the control of speech and the press, that is generally thought relevant to knowledge. The discussion of how power is and should be used to affect what we *do* is not thought particularly relevant to an epistemological understanding of science.

These considerations lead to the final point, that power is essentially repressive. Power uses punishments and rewards to prohibit, censor, constrain, and coerce. Hubert Dreyfus and Paul Rabinow describe the received understanding of power as

> a tradition which sees power only as constraint, negativity, and coercion. As a systematic refusal to accept reality, as a repressive instrument, as a ban on truth, the forces of power prevent or at least distort the formation of knowledge. Power does this by suppressing desire, fostering false consciousness, promoting ignorance, and using a host of other dodges. Since it fears the truth, power must suppress it.[39]

The condemnation of Galileo by the church and the suppression of Mendelian genetics in the Soviet Union have served as paradigms of this aspect of the encounter between power and knowledge. Power can be used against persons who espouse dangerous ideas or restrained to favor those whose ideas are more in keeping with the interests or preferences of those who wield power. But it cannot create new ideas or new knowledge. Its operation presupposes such achievements, which it can then protect or suppress. Once again, the

37. See, for example, Geuss 1982: 12–22.
38. Laudan 1977: 4; quoted above, p. 1.
39. Dreyfus and Rabinow 1983: 129.

externality of power and knowledge is considered central: power can be used to suppress or encourage the achievement of knowledge, but that achievement itself is constituted without reference to power. Power operates *upon* science, for example, but does not operate within it.

Empirically, of course, this claim is clearly false. Scientists employ power in the course of their activities just as do persons pursuing other activities in other institutions. Questions of political influence, career development, financial constraint, legal prohibition, ideological distortion, and so forth also arise in science, which after all is not wholly isolated from worldly considerations. This embarrassing intrusion of mundane concerns and pressures is usually circumscribed philosophically, however, by emphasizing a *conceptual* separation between science viewed as a field of knowledge and science viewed as a field of power. Various attempts to distinguish between internal and external history of science, or between philosophy and sociology of science, reflect this desire to sustain a conceptual separation between science and the way power operates within or upon it. What is "internal" to science are the cognitive, rational, intellectual, and epistemological concerns and activities that account for the development of knowledge. The effects of political, sociological, and individual psychological factors upon the development of science are external to knowledge and can be compartmentalized in separate inquiries. A complete picture of science as it actually is requires both, of course, but only the former is necessary to understand what is essential to or characteristic of science as a rational enterprise.[40]

Toward an Alternative Account of Power and Knowledge

The central thesis of this book is that the preceding account of the relations between political power and scientific knowledge is seriously misleading. It leads us to overlook important ways power is exercised today and to misunderstand both scientific practices and their political effects. Taken together, the pragmatist and new empiricist developments in the philosophy of science can help us glimpse these inadequacies in the received view of the relations between power and knowledge.

The first, partial challenge to the received understanding of power comes from the pragmatists. They need not and do not dispute the view that power is a commodity agents possess and exercise, that it is

40. As examples see Laudan 1977: 196–222; Lakatos 1978: 102–38.

essentially repressive, or that it operates primarily on and through beliefs. But they do challenge the claim that it is external to knowledge. Pragmatism rejects the idea that there are identifiable criteria for truth apart from what we arrive at through the practice of inquiry. Everyone agrees that the empirical claims of science can be identified only through our inquiries; pragmatists argue also that the standards and criteria by which we appraise our results are themselves a product of inquiry and cannot be identified in advance. What *counts* as an acceptable conclusion, and what counts as a good reason for its acceptance, also changes as inquiry proceeds.[41]

The importance of this approach is that it undermines the distinction that permitted us to insulate knowledge from the operations of power. The way the world is can no longer be so clearly distinguished from the social context within which it is disclosed. If what is true could be (in principle) identified independent of the course of any particular tradition of inquiry, then it might make sense to distinguish the way power affects what we believe from truth's imperviousness to power. But if the criteria for what can count as true are themselves products of inquiry, then an understanding of power must be integral to epistemology rather than just a by-product of its exclusions. The exclusion of the distorting influences of power becomes not just a desideratum for the actual achievement of knowledge but a consideration for the theory of knowledge as well. Gary Gutting illustrates this point nicely in a discussion of Habermas:

> The connection between epistemology and social and political philosophy is perhaps most readily seen by reflecting on the implications of a pragmatic theory of truth. Truth, it is said, is to be understood in terms of the ultimate consensus of the community of inquirers. . . . [But] even such an "ideal" (indefinitely distant) consensus is not sufficient. For inquiry might continue indefinitely under conditions of psychological or political repression that would distort its conclusions at every stage (e.g., yield ideology rather than science). Accordingly, the possibility of truth—and hence of knowledge—requires the realization of a social and political environment that will permit inquiry without distorting constraints.[42]

Proponents of pragmatism have been concerned primarily with the relation between things and the ways they are disclosed, arguing that what things are depends partly upon how they show themselves. What

41. Laudan 1977: 129; Hacking 1983: 127–28; Putnam 1978: 134–37.
42. Gutting 1979: 104.

there is depends above all on what languages or vocabularies are available to describe what there is and on what values or "coherence conditions" govern the adequacy of our descriptions.[43] What we can now recognize is that this view partially transforms the epistemological problem of distinguishing true from false (warranted from unwarranted, or rational from irrational) beliefs into the political problem of distinguishing free inquiry from inquiry constrained and distorted by the exercise of power. A political concern for whether scientific discourse is free and undistorted therefore becomes an essential part of any reflection on the cognitive aspects of science.

Proponents of the new empiricism have not extensively discussed the implications of their views for our understanding of power and its relation to knowledge. When they have mentioned it, they have usually affirmed traditional accounts unquestioningly.[44] Nevertheless, if we take the new empiricism seriously, it forces us to reappraise the relation between power and knowledge in a more radical way. The central issue is no longer how scientific claims can be distorted or suppressed by polemic, propaganda, or ideology. Rather, we must look at what was earlier described as the achievement of power through the application of knowledge. But the new empiricism also challenges the adequacy of this description in terms of "application." The received view distinguishes the achievement of knowledge from its subsequent application, from which this kind of power is supposed to derive. New empiricist accounts of science make this distinction less tenable by shifting the locus of knowledge from accurate representation to successful manipulation and control of events. Power is no longer external to knowledge or opposed to it; power itself becomes the mark of knowledge.

Why is this so? Generally speaking, the new empiricists give antirealist interpretations of scientific theories: successful theories need not accurately represent what goes on in the world (Nancy Cartwright even argues, in the case of "fundamental" or explanatory theories, that they cannot do so if they are to be successful).[45] Nevertheless, the new empiricists tend to be realists about the *entities* theories refer to. Science need not increase the accuracy of our representation of things, but it must improve our ability to cope with them. The increasing effectiveness of our technical capabilities, and the increasing reliability of our predictions, characterizes scientific knowledge, regardless of whether

43. See, for example, Putnam 1978: 127–39; Rorty 1979: 315–42; Hesse 1980: 125–38.

44. Laudan 1977: 4, 141; Hesse 1980: 196–97.

45. Cartwright 1983: 44–53.

our theories display increasing verisimilitude (whatever that would be). In Hesse's terms, the pragmatic criterion measures the success of science independent of any assessment of the representational accuracy of theories. This view makes untenable the traditional account, that technical control is a result of theoretical knowledge. Technical control, the *power*[46] to intervene in and manipulate natural events, is not the application of antecedent knowledge but the form scientific knowledge now predominantly takes. The new empiricism therefore needs a more detailed account of how such power/knowledge operates in the sciences.

Note that the technical control or pragmatic success claimed to be characteristic of science is not the same as the development of socially useful technologies. The two have important relations to one another, but they must be kept distinct. Some phenomena are brought under increasingly precise and reliable control in the course of scientific research without yielding any technological benefit; such research is technically sophisticated but technologically irrelevant. Some technologies do result directly from basic research, but the considerable developmental work the technology usually requires may be of little direct scientific interest. And of course, some technologies are developed with little significant direct input from scientific research. Unfortunately, this distinction between science and technology has too often been misunderstood, so that the achievement of technical control in the laboratory is overlooked or underestimated. If we are to understand what role the sciences do play in the development of socially useful (or harmful) technologies, we need a better grasp of the technical aspects of science itself.

It is not difficult to point out some fundamental contrasts between the analysis of power/knowledge that the new empiricism needs and the received view of the relation between power and knowledge as I have outlined it. Most obviously, the understanding of power as essentially repressive or censorious is inappropriate here. The power characteristic of scientific knowledge (at least in the natural sciences) does not operate directly upon or against persons and their beliefs. It is a constructive power that reshapes the world and the way it is manifested; it is of "nature," not persons, that Bacon proposes to "bind her to your service and make her your slave."[47] This claim does not imply that the construction of phenomena and the reconstruction

46. The use of the term "power" to describe these technical capabilities is not accidental here, nor is it just equivocation; in chapter 7 I will argue that these technical capabilities are closely integrated with more obviously "political" power relations.

47. Farrington 1951: 197. My attention was drawn to this passage by Keller (1985).

of the world are altogether desirable, or that they do not indirectly deploy power against persons. Consider, for example, feminist criticisms of such an approach to science.[48] Such questions must be held in abeyance until chapter 7. But I can say that this mode of operation of power is productive rather than repressive. Knowledge arises out of the development and exercise of this sort of power rather than only being suppressed or permitted.

Although theories and other beliefs about how the world is clearly play an essential role in science, the new empiricism challenges traditional accounts of that role. The technical power of science results from something other than just the representational accuracy of scientific theories. The account I shall develop will focus upon scientific skills, practices, and techniques as the place where this kind of power is developed and operates. This approach confronts a tradition going back to Plato, in which the only alternative to theoretical knowledge of what is "truly real" is a blind knack that cannot be taught or understood. To understand scientific knowledge, we need instead a positive account of the skills and practical know-how that construct and stabilize phenomena and that enable scientists to intervene and to manipulate them in informative ways. Only in such skills and practices do knowledge and power come together. The relation between practices and theories is complex, however; it is not a straightforward opposition of the two, since some of the practices I shall discuss are strategies for representing and theorizing. This issue will be discussed in chapter 4. The important point in the meantime is that the hypothetico-deductive model of science's predictive success needs an alternative. Science's technical capabilities (powers) are developed, communicated, and preserved without necessarily being mediated by theoretical representation. We need a detailed understanding of how this process occurs.

This point leads to a related issue, which I will call the decentralization of scientific power/knowledge. The standard model of scientific knowledge takes it to be knowledge of universal laws, valid at all times and places. These laws can be applied to particular situations, however, only by using various bridge principles and by determining the relevant facts about the situation that need to be included in this instantiation of the law. Thus the problem is always how to bring universally valid knowledge to bear on local situations. The ability to manipulate and control events in the laboratory derives from knowledge (or hypothetical assumption) of universal laws and theories. The

48. Cf. Keller 1985; Merchant 1980; Harding and Hintikka 1983.

new empiricism suggests an analysis moving in the opposite direction. In scientific research, we obtain a practical mastery of locally situated phenomena. The problem is how to standardize and generalize that achievement so that it is replicable in different local contexts.[49] We must try to understand how scientists get from one local knowledge to another rather than from universal knowledge to its local instantiation. I do not deny, absurdly, that theories and laws play a vital role in the development and transfer of scientific knowledge; I only say that a new account of that role is called for. Theories and laws are understood in and through concrete cases; abstract formalisms get their sense from particular uses, which must then be "translated" into possibilities for replicating or transforming those uses in different contexts. Only with such an account, I will argue, can we defend the new empiricist insight that experiment has a life of its own and that the pragmatic success of science is not prevented by changing theoretical interpretations of that success.

It is important to note from the outset, however, that I am not advocating a shift from deductivist to inductivist models of scientific inference. Scientific research to a large extent involves producing reliable and repeatable effects that can pose a significant scientific problem or help solve one. The difficulties one confronts locally in this attempt—in designing and setting up the research, getting the equipment to work reliably, and interpreting the results—present problems specific to this occasion with its local constraints and resources.[50] Even the interpretation of the problem, and of what has been achieved in its solution, has its locally idiosyncratic aspects.[51] The issue does not concern inferring a general claim from a number of claims about particulars: instead it concerns simplifying and generalizing a problem formulation, replicating an achievement in different circumstances or for different purposes, or adapting a result to help understand a new situation or problem or to construct new possibilities for investigation. In short, it is a matter of transforming a concrete, local achievement in order to open up a field of possibilities for further scientific investigation. Philosophers have generally assumed that science can be understood as the production of a network of statements; the new empiricism forces us to confront the possibility that it can be better interpreted as an interrelated field of activities and achievements. The question is not how we infer one statement

49. Ravetz 1971: 197–202; Latour 1983.
50. Knorr-Cetina 1981: 9, 33–44; Latour 1983: 146–53.
51. Ravetz 1971: 192–94.

from others, but how we generate research opportunities and resolve them, drawing upon and reinterpreting prior concrete achievements.

One might say that the traditional philosophical model of the local site of research is the observatory. Scientists look out at the world or bring parts of it inside to observe them. Whatever manipulative activities they perform either are directed at their instruments or are attempts to *re*produce phenomena in a location and setting where they will be observable. The new empiricism leads us instead to take seriously the *labor*atory.[52] Scientists produce phenomena: many of the things they study are not "natural" events but are very much the result of artifice. The classical sciences were astronomy, mathematics, and mechanics and descriptive biology, anatomy, and geology, whose objects were already available for description and reflection. More characteristic of science as it is practiced today are high-energy physics, electromagnetism, thermodynamics, chemistry, and experimental biology in its many subfields. Many, if not most, of their objects of study do not appear in nature in the form in which they are investigated. Thermodynamics studies the transformation of heat into mechanical work; it is the science of the heat engine. Electrical currents and their myriad effects are equally the product of artifice; they did not exist in significant measure before the nineteenth century. The "particles" studied in high-energy physics research require massive, complex apparatus to produce them as well as to register their presence and their behavior. In chemistry it is not just that the vast majority of substances chemists study do not occur except as the products of manufacture; even the most ordinary chemical materials appear in standardized, highly purified form. Likewise, biologists study cultured cells, bacteria cloned for thousands of generations in laboratories, germ-free strains of animals, and a host of other "nonnatural" objects. Nor have the classical sciences themselves remained free from this "Baconian" tendency. The study of artificially constructed or manipulated phenomena has become characteristic of most of the modern sciences.

Two essential points can be noted about this artificiality. First, objects must be significantly transformed in order to be reliably manipulated and informatively understood in the laboratory. Science must work extensively on its objects before it can work with them. Second, these transformed objects are often then built into the world outside the laboratory. The "natural" world itself is now to an astonishing extent the product of artifice. These points can suggest two important

52. See Latour 1983: 159–69.

directions for studies of the relations between science and technology, both of which will be important for an inquiry into the relations between knowledge and power. We must ask how and to what extent technological development more or less independent of science provides science with its objects and materials. But we must also ask whether the influence of science upon technology is at least as much the transformation of scientific processes, techniques, and practices to satisfy extrascientific concerns as it is the application of scientific theories. Perhaps this transformation of scientific practices is not different in kind from the transformations that I have suggested take place between scientific contexts.

The account of knowledge and power I will attempt to develop in what follows will draw upon both of these challenges to the tradition. From the new empiricists, I take the implicit recognition that power does not merely impinge on science and scientific knowledge from without. Power relations permeate the most ordinary activities in scientific research. Scientific knowledge arises out of these power relations rather than in opposition to them. Knowledge is power, and power knowledge. Knowledge is embedded in our research practices rather than being fully abstractable in representational theories. Theories are to be understood in their uses, not in their static correspondence (or noncorrespondence) with the world. Power as it is produced in science is not the possession of particular agents and does not necessarily serve particular interests. Power relations constitute the world in which we find particular agents and interests. As Foucault has remarked, "Power must be analyzed as something which circulates, or rather as something which only functions in the form of a chain. It is never localised here or there, never in anybody's hands, never appropriated as a commodity or piece of wealth. Power is employed and exercised through a net-like organization."[53]

This assertion does not mean that power as it has traditionally been discussed—repressive juridical power—no longer exists or does not affect science. But it does mean that we cannot ignore or discount these "capillary" power relations and that the influence upon the sciences of the more traditionally recognized forms of power must be understood against the background of the capillary power relations that traverse scientific practices themselves.

From the proponents of pragmatism, I draw the insight that power and knowledge or truth are internally related. The power relations that open up a field of scientific practice are also relations of dis-

53. Foucault 1980: 98.

closure, of truth. In working on the world, we find out what it is like. The world is not something inaccessible on the far side of our theories and observations. It is what shows up in our practices, what resists or accommodates us as we try to act upon it. Scientific research, along with the other things we do, transforms the world and the ways it can make itself known. We know it not as subjects representing to ourselves the objects before us, but as agents grasping and seizing upon the possibilities among which we find ourselves. The turn from representation to manipulation, from knowing that to knowing how, does not reject the commonsense view that science helps disclose the world around us.

Chapter 2

Science as Practice: Two Readings of Thomas Kuhn

THE most influential attempt to consider science as a field of practices rather than a network of statements, Thomas Kuhn's *The Structure of Scientific Revolutions,*[1] has also been perhaps the most misunderstood. In particular, the depth of his criticism of representationalist epistemology has often been overlooked. Kuhn has most commonly been read by philosophers as someone who ascribes a leading role to theory in science, who emphasizes the noncumulative character of theory change, and who denies the possibility of nonneutral criteria for assessing the cognitive worth of such changes. Kuhn has objected vigorously to such interpretations. At one point, he even mused:

> I am tempted to posit the existence of two Thomas Kuhns. $Kuhn_1$ published in 1962 a book called *The Structure of Scientific Revolutions.* . . . $Kuhn_2$ is the author of another book with the same title. . . . That both books bear the same title cannot be altogether accidental, for the views they present often overlap and are, in any case, expressed in the same words. But their central concerns are, I conclude, usually very different. As reported by his critics, $Kuhn_2$ seems on occasion to make points that subvert essential aspects of the position outlined by his namesake.[2]

I propose to extend Kuhn's Borgesian fantasy of two Kuhns a little further in order to see what is at issue in the various readings of his book. One of the issues that underlies the plasticity of his work but has

1. Kuhn 1970a.
2. Kuhn 1970b: 231.

not been adequately discussed is the contrast between a representationalist, theory-dominant account of science and one in which the practices of scientific research take precedence. My account of the critics' Kuhn, Kuhn_2, will attempt to reproduce common themes found in the influential responses to Kuhn by Israel Scheffler, Dudley Shapere, Abner Shimony, Frederick Suppe, and others.[3] These interpreters all are concerned to rebut what they take to be Kuhn's radical criticism of traditional views of science. If I am right, they miss important aspects of his work because they amalgamate it too thoroughly with the representationalist tradition they defend. I base my interpretation of Kuhn_1 primarily upon Kuhn's attempts to distinguish his position from his critics' interpretation.[4] Nevertheless, I undoubtedly take him further in the direction of an account of science as practice than he himself would be happy with. I do so in order to sharpen the contrast between the two "Kuhns," but more importantly because I believe this provides an illuminating reading of his book, one that is generally consistent with the text but that has to my knowledge not been seriously discussed. Any attempt to understand Kuhn's own position must ultimately take account of both readings, for his critics have not completely misunderstood him. Kuhn has been strongly influenced by the epistemological tradition he challenges. I will not try to make such a reconciliation here. My aim is not commentary, but the development of an interpretation of science whose roots in Kuhn are too often unnoticed.

The Philosophers' Kuhn

Let us begin with Kuhn_2, Kuhn as he has been read by many of his philosophical critics. Kuhn_2's analysis of science begins with an account of "normal science," wherein scientists take an uncritical attitude toward the most fundamental theories and concepts accepted in their field. These fundamental concepts and theories are embodied in a paradigm, a set of theoretical doctrines constituting a worldview, which was postulated in the work that originally established the field of research (or reestablished it on a revolutionary basis). This paradigmatic worldview serves a number of important functions for those who engage in normal science. It prescribes some beliefs as essential and proscribes others. It determines which facts it is important to

3. Shapere 1964, 1971; Scheffler 1967, 1972; Suppe 1977: 135–51, 636–49; Shimony 1976.

4. Kuhn 1970a: 174–210, 1977: 266–339.

know and provides some general expectations of what those facts will be. These expectations will not always be met, however. Such failures, or anomalies, provide the puzzles that keep normal science occupied. Normal science is essentially puzzle solving, attempting to reduce the discrepancies between a paradigmatic worldview and the world and to fill in the many blanks left open by the original sketchy development of the worldview. The distinctive feature of such puzzle solving is that it cannot challenge the fundamental beliefs taken from the paradigm. If a scientist fails to reconcile theory and evidence, it represents a failure of the scientist, not the theory. A paradigm guides puzzle solving in three ways: its values determine which puzzles are worth solving and what those solutions are supposed to achieve; the paradigm specifies the standards for acceptable solutions to puzzles; and finally, paradigms suggest model problem solutions (exemplars) that heuristically guide scientists toward such solutions.

Paradigms are closely linked to the scientific communities that accept them. Neither paradigms nor scientific communities can be fully identified apart from one another. A scientific community consists of those scientists who accept the same paradigm and regard its theoretical doctrines as inviolable. Conversely, a paradigm is the set of inviolable theoretical doctrines accepted by members of a scientific community. Those who do not fully accept the key elements of the paradigm are read out of the community. Their work is not taken seriously, and their criticisms are listened to only politely, if at all. As a result, normal science proceeds without controversy over fundamental principles, enabling scientists to communicate through a specialized literature that takes these principles for granted and to investigate increasingly esoteric phenomena with highly sophisticated equipment. Scientists' efforts are devoted to getting on with such specialized research rather than to quarreling over the most basic theoretical issues.

All paradigms confront counterinstances (anomalies) at all times. Initially they have solved only a few problems, and their appeal depends upon promissory notes. As they become more refined and solve more problems, new anomalies appear because the paradigm can be compared with the world more precisely, at more points. Those paradigms that cease to conflict with the world no longer provide material for research and therefore cease to sustain a scientific community. Outstanding anomalies are generally treated as puzzles that one can expect to solve with enough ingenuity. But sometimes, when anomalies become sufficiently numerous, persistent, or weighty (or their appearance accelerates), community consensus over the paradigm begins to erode. This situation can occur especially if an important puzzle repeatedly resists solution by the ablest and most prominent scientists in the field. This communal loss of confidence causes (or allows) scientists to tinker with fundamental elements of the paradigm, in the hope of finding an unorthodox answer to puzzles for which they have come to despair of a solution. In such periods of crisis, the intelligibility of work within the field is increasingly threatened. No one is quite sure anymore what beliefs can be taken for granted, what projects are worth undertaking, or what counts as a legitimate solution to a problem. Crisis leaves a scientific community in frustrating disarray.

Crises become especially interesting philosophically when an unorthodox solution to an outstanding puzzle is proposed. (Sometimes, of course, the solution was proposed much earlier, but was taken seriously only when the crisis developed.) The community then polarizes into proponents of the paradigm embodied in this new proposal and defenders of the old orthodoxy. Because paradigms inculcate the most fundamental principles, vocabulary, standards, and values, these disputes cannot easily be resolved. The competing paradigms are incommensurable. Their terms have different meanings and may have different referents. There is no neutral evidence with which they can be compared, because they lead their proponents to see the world differently and to describe it in different terms. They differ on what counts as a solution to a scientific problem, and they often endorse, or at least emphasize, different values regarding what science should achieve. Because of this incommensurability, scientists arguing for different paradigms often "talk through one another," failing to grasp fully their opponents' arguments or sometimes even their conclusions. There is no place for logical demonstration or proof here. Arguments can at most be persuasive. The switch to a new paradigm can be likened to a religious conversion or a Gestalt switch, in which scales drop from one's eyes so one suddenly sees the world differently and may be unable to recreate one's earlier perspective. As one side comes to predominate in these struggles, its opponents are either converted in turn or overpowered. Their work is ignored and their protests go unheeded as their colleagues close ranks around a new normal scientific consensus. Thus go scientific revolutions.

Science, according to Kuhn_2's account, develops in a cyclical pattern of normal science, crisis, revolution, normal science. If we look back upon this pattern from the viewpoint of the victors, scientific revolutions inevitably appear progressive. The problems affecting the old paradigm were resolved, and a new basis was established for proceeding with research. But this revolution constitutes progress only if we agree that the problems set by the new paradigm are genuine and important and that the solutions proposed for them are legitimate. If we join the "reactionaries" who challenge this view, or even if we merely agree that the point is at issue and cannot be presupposed, progress cannot be so easily discerned. There is no paradigm-neutral standard with respect to which progress can be assessed. On $Kuhn_2$'s account, the scientific revolutions that have led us to our current beliefs may easily reflect the triumph of "mob psychology" rather than a march toward truth.

The Radical Kuhn

$Kuhn_1$ uses many of the same terms to describe the development of science, but the concepts embodied in these terms are rather different. According to $Kuhn_1$, normal science is research in which scientists know their way about. Through training and experience they have a reliable sense of what they are dealing with, what can affect it, how it can make itself known, and what they can do with it. This confidence is developed from their practical grasp of one or more paradigms, concrete scientific achievements that disclose a field of possible research activities. Paradigms are not primarily agreed-upon theoretical commitments but exemplary ways of conceptualizing and intervening in particular empirical contexts. Accepting a paradigm is more like acquiring and applying a skill than like understanding and believing a statement. Actually, it involves multiple skills simultaneously: applying concepts, employing mathematical techniques (not just calculating, but choosing the right mathematics, applying it correctly to an empirical situation, knowing its limitations and approximations, etc.), using instrumentation and other apparatus, and recognizing opportunities for varying or intervening in particular theoretical or experimental situations. Kuhn emphasizes the role of analogies in grasping a paradigm: an essential skill is understanding how to treat new situations like old ones, to do for them what has already been done in the exemplary case.

Normal science is thus characterized by the *use* of the same paradigms. Sharing a paradigm leads to other common features found in the practice of normal science: shared concepts, symbolic generalizations, experimental and mathematical techniques, even theoretical claims. But these other features may be more loosely agreed upon, and the agreement itself may be more problematic. Kuhn insists that scientists can "agree in their *identification* of a paradigm without agreeing on, or even attempting to produce, a full *interpretation* or *rationalization* of it. Lack of a standard interpretation or of an agreed reduction to rules will not prevent a paradigm from guiding research."[5] Normal science involves shared practices, not shared beliefs. These shared practices take place within a common field of operations, which Kuhn has more recently called a "disciplinary matrix." A disciplinary matrix is the "field" within which the shared concepts, symbols, apparatus, and theories are applied. It opens up a domain of objects for comprehension, manipulation, and intervention. This domain constitutes a field of research possibilities and opportunities arising out of prior activities and achievements. There is room for considerable disagreement within such a field, often about very basic issues. What is shared, however, is a sense of what is at issue, why it matters, and what must be done to resolve it. Such agreement is not merely compatible with disagreements about specific issues within the field; it is what makes significant disagreement intelligible. Kuhn emphasizes this in a retrospective account of his work.

> I [once] conceived normal science as a result of a consensus among the members of a scientific community . . . in order to account for the way they did research and, especially, for the unanimity with which they ordinarily evaluated the research done by others. . . . What I finally realized . . . was that no consensus of quite that kind was required. . . . If [scientists] accepted a sufficient set of standard [ways to solve selected problems], they could model their own subsequent research on them without needing to agree about which set of characteristics of these examples made them standard, justified their acceptance.[6]

The research activities of normal science can be usefully compared to puzzle solving in that they are well-defined tasks with what are presumed to be determinate solutions that require ingenuity to find. But there is no single characterization of what these puzzles are about, such as Kuhn$_2$'s account of them as discrepancies between theory and evidence. Kuhn discusses a number of types of such puzzles. The first is the determination (or more precise determination) of facts that show up as important within the disciplinary matrix. Similar problems arise in developing and standardizing techniques for investigating particular objects or situations within the matrix. A second kind of problem consists of experimentally exhibiting the contact between the theories that stem from the paradigm and relevant empirical phe-

5. Kuhn 1970a: 44.
6. Kuhn 1977: xviii–xix.

nomena. This is largely a problem of getting theory-governed relationships to show themselves clearly. Such touchpoints between theory and world "test" the theory, but only in a limited way. They do not determine whether the theory succeeds or fails *tout court* but at most show whether, where, and to what extent the theory needs articulation or refinement. The question at issue here is how precise it is, not whether it is correct. Finally, a medley of activities are grouped under the heading "paradigm articulation." They include determining physical constants, developing lawlike generalizations, extending paradigm solutions to other phenomena, reformulating paradigmatic theories and concepts, and so forth. This category includes resolving a variety of anomalies and conceptual conflicts, but the principal aim is to extend the scope and power of one's know-how rather than to remove objections to one's theories. Anomalies are not normally objections according to Kuhn_1's account, because science is more an activity of theory development than of theory appraisal.

Kuhn_2 describes scientific groups as communities of believers. For Kuhn_1 they are communities of fellow practitioners. The former account insists that such communities cannot tolerate fundamental disagreement. The latter insists that scientific communities are rife with fundamental disagreements but rarely exhibit controversy over fundamentals because research can proceed coherently and intelligibly without resolving all the disagreements. Shared paradigms enable scientists to identify what is worth doing, and what has or has not actually been done, without agreeing on how to describe this body of material. Scientists are ignored, or read out of the community, not for disagreeing with others but for doing work that does not fit in with what others are up to. In evaluating others' work (even in deciding whether to read it) scientists ask, Can I make sense of this? Can I or do I need to take account of it in my own work? Is it reliable? Scientists can hold unusual or unpopular beliefs about issues in the field, so long as they do research that others can take account of and use in their own terms. What matters is that their *work* is relevant and reliable in the judgment of others. Scientific communities are characterized by common problems and techniques and by reference to the same achievements, not by monolithic consensus.

All paradigms confront obstacles (anomalies) at all times. Anomalies play an important role in Kuhn_1's account, but it is not that of counterinstances to received theories. To be aware of a counterinstance is to know what you are dealing with and how it stands in relation to your previous grasp of things. The recognition of anomalies is instead an awareness that something significant is not understood or not being dealt with adequately, but it is not yet a clear awareness of what the problem is. "Assimilating a new sort of fact demands a more than additive adjustment of theory, and until that adjustment is completed . . . the new fact is not quite a scientific fact at all."[7] Until one better understands where to locate this anomalous fact in one's scheme of things, one is not sure what it signifies, what can be done with it, in short, what it is. How scientists respond to such ambiguous difficulties often depends upon whether the problems they present seem localizable. An anomaly that does not show up in other contexts and does not seem closely connected with objects or techniques one regularly employs can easily be dismissed as an artifact. But if it seems to affect other things one might want to do, it must be dealt with. Kuhn cites the example of Roentgen's discovery of a glowing barium platinocyanide screen in the vicinity of his shielded cathode-ray tube.[8] This was important and needed to be investigated, because cathode-ray tubes were important research tools. The unexpected glowing screen suggested that Roentgen did not fully grasp what was going on in his research apparatus. Until the scope of that failure was clarified, further ordinary research using cathode-ray tubes would have been pointless. Dealing with anomalies sometimes requires revisions in theory in order to assimilate and understand a new phenomenon. At other times it requires revised techniques or instruments that circumvent the anomalous effect without necessarily explaining it. This situation occurs because, to repeat, anomalies are not conflicts with theory but practical difficulties (which sometimes become opportunities). They need to be resolved to the extent necessary to get on with research. If an anomaly is sufficiently obstructive, or interesting in its own right, it can replace the original research topic. More often, perhaps, it is a "pitfall" to be negotiated around so one can proceed securely with one's original project.[9]

Most anomalies are either resolved or bypassed fairly quickly, but some persist and resist all efforts to assimilate them. If their import is not clearly localizable (so that they can be ignored by most scientists), the result will be highly disconcerting. It means that not just the anomaly, but whatever causes it and whichever situations it occurs in, are not reliably understood. At its worst the ensuing crisis places in doubt the intelligibility and reliability of many research practices and achievements. Once again, it is not that scientists do not know what to

7. Kuhn 1970a: 53.
8. Kuhn 1970a: 57.
9. Ravetz 1971: 94–101.

believe; scientists are professionally accustomed to uncertainty of *that* sort. It is that they are no longer quite sure how to proceed: What investigations are worth undertaking, which supposed facts are unreliable artifacts, what concepts or models are useful guides for their theoretical or experimental manipulations? Crisis is never total, or research would collapse into total unintelligibility: there would no longer be a shared field of activity. But crisis expands and blurs the bounds of that field and thereby blurs one's sense of the place of one's activities within the field. It makes sense to try different things, but what sense those things make is no longer quite so clear.

As for Kuhn_2, crisis culminates in the proposal of an alternative paradigm. For Kuhn_1, however, what is proposed is not a theoretical worldview but some concrete achievement put forward as a new focus (and model) for research. New concepts and theories will undoubtedly emerge from this achievement, yet what these are may be disputed even by its proponents. The prior dispute, between proponents of different paradigms, concerns how science should be practiced: what should be done, what must be taken account of, and what counts as a significant and reliable result. Such disputes can be difficult to resolve, because the protagonists work in different "worlds." Their research "fields" are organized differently. If they have difficulty communicating fully, it is not because they fail to understand or properly translate one another's words and sentences (although this can happen on occasion). Instead, they may not fully grasp the point of doing the sorts of things the others do. Changing from one paradigm to another is not like a conversion to new beliefs but is like a conversion to a new form of life. Such conversions in science are usually accompanied by extensive reasoning and argument. But this cannot be conclusive by Kuhn_1's account, because the *force* of these arguments is fully apparent only to those already at home in the new disciplinary field.

Kuhn_1 does not impose a rigid schematism of normal science–crisis–revolution–normal science upon the history of science. Normal science and crisis are not historical periods but ways of practicing science. One or the other may predominate at any one time in a discipline, but this predominance is not essential (Kuhn does think this state of affairs is in fact common). Some scientists may experience crisis over problems that do not disturb their colleagues. Others blithely go on with their normal research even though their peers are no longer quite sure what to make of it. Normal science and crisis almost always coexist in this way, and how to categorize the dominant trend is a matter of historical judgment.

Revolutions are likewise open to interpretation. How revolutionary a new development is depends in part upon how one interpreted the paradigms preceding it; some interpretations make it more wrenching than others. This explains how Einstein could say that special relativity simply worked out the implications of Maxwell's dynamics, whereas most commentators took relativity to be a revolutionary reconstruction of mechanics. His interpretation had already moved Maxwell's widely accepted theory far from classical mechanics, partially closing the conceptual gap supposed to separate Maxwell's work from his own. His interpretation thereby transformed the development of special relativity into something more like paradigm articulation than like radical revolution (note that this reading in no way reduces the significance of Einstein's achievement, because being able to see these implications in Maxwell and point them out to others is just as noteworthy as developing them without antecedents). The ambiguity between revolution and normal science is reinforced by the fact that many scientific practices are retained (perhaps in reinterpreted forms) across revolutions. The interpretive question is whether to emphasize the continuities or the discontinuities. In any case, the border between revolution and a significant development within an evolving tradition of normal science need not be a sharp one.

Philosophical readers of Kuhn have usually equated scientific revolutions with major conceptual and theoretical changes in a field. We can see, however, that this association need not be the case for $Kuhn_1$. New instruments, techniques, or phenomena can cause equally fundamental changes in the way research is done within a given field. A good example is the development of recombinant DNA techniques in biology. These techniques dramatically changed the questions one could ask, and the kinds of answers one could expect, within some areas of molecular genetics. The result was a rapid change in the kinds of research it made sense to do in those fields. The revolution in high-energy physics in the late 1960s provides another case in point. This revolution is often described in terms of the use of gauge field theories and the adoption of quark models for interpreting hadrons. But gauge theory dates back to 1954, and the first quark models were proposed in 1964. The significant change throughout high-energy physics as a whole may be represented more effectively by the shift in experimental work (and correlative changes in the phenomena that interested theorists) in the late 1960s and early 1970s. The introduction of electron-positron colliders, and a new emphasis upon lepton beams rather than hadrons and upon hard rather than soft scattering of hadron beams, marks a decisive shift in the field. Inevitably, theoretical and conceptual changes were involved as well. But the revolution can be seen as focused upon the introduction of a new instrument and some different uses of old instruments.[10] Still a third example is offered by the "phage group" in biology in the 1940s. Their work revolutionized genetics. They are often noted for bringing the intellectual style and interests of physicists into biology. It is no accident, however, that the innovation that gave the group its name was a change in the organisms investigated by geneticists, from eukaryotes to bacteriophages. Their theoretical innovations followed from this change in focus rather than initiating it.[11]

Whatever their origin, revolutions represent progress from a researcher's point of view (sometimes even recognized by their opponents) because they allow research to proceed coherently again. Modern chemistry thus represents progress over phlogiston chemistry, because we know what to do with the former but not the latter. For a scientist, this lack of knowledge must represent a decisive objection to phlogiston theory. Paradigms can be criticized on other grounds, however. They may be more or less elegant, offer guidance to other disciplines or not, cohere or conflict with our social and political practices, and so forth. Unless we assess them with reference to these other issues, however, we do not know how to ascertain progress with respect to truth. Getting research to progress is of intense interest to scientists; whether such progress takes place may or may not seriously concern others. Which forms of progress really matter may be hotly disputed in a variety of forums. But there is no such thing as scientific progress *simpliciter,* except as the result of an exercise of judgment balancing these various ways progress might be assessed.

Conclusion

What are we to make of these two ways of reading Kuhn? The first reading, Kuhn_2, treats science as the construction and appraisal of theories that aim to represent the world. It is replete with words like "believe," "accept," "see" or "observe," "theory," "counterinstance." Kuhn_2 challenges earlier philosophical accounts of what is involved in justifying and accepting a theory or observing the outcome of experiments, but he does not deny that these are the important issues in the

10. For a detailed account of these developments, see Pickering 1984.
11. See Allen 1975, chap. 7.

philosophy of science. Kuhn$_2$'s principal philosophical innovations were the following seven points:[12]

1. Science always requires some theoretical presuppositions that cannot be independently justified.
2. There is no theory-neutral observation language, nor do theories consist in models external to the facts they cover; theories constitute the way we observe and describe the facts.
3. There have been radical changes over time in both scientific theory and observations; scientific knowledge is noncumulative.
4. The language of scientific theories is not precise.
5. The meanings of scientific concepts are determined at least in part by their place in theoretical networks, not by their correspondence with independently observed facts.
6. Theories are not tested directly by empirical evidence; they are evaluated only by comparison with rival theories; furthermore, because theories are logically incommensurable, even these comparisons cannot be conclusive; there may be no rational, noncircular way to assess and compare theories conclusively according to their empirical adequacy.
7. The contexts of discovery and justification cannot be clearly distinguished; an essential aspect of the justification of theories is their fruitfulness in leading to new discoveries; an essential aspect of discovery is the justification of the discovery as genuine, including defending the theoretical context within which it appears.

These are certainly important and interesting claims; all of them have been seriously defended by some philosophers, and some are so widely accepted as to begin to constitute a new orthodoxy in the philosophy of science. If they are not what Kuhn intended, one can certainly sympathize with Alan Musgrave's assessment of Kuhn's retrospective self-interpretations: "Kuhn's *Postscript* left me feeling a little disappointed. I find the new, more real Kuhn who emerges in it but a pale reflection of the old, revolutionary Kuhn. Perhaps this revolutionary never really existed—but then it was necessary to invent him."[13]

Musgrave might, however, be less disappointed, and perhaps once again dismayed, to encounter Kuhn$_1$ as I have presented him here.

12. I have been influenced by similar lists in Hesse (1980: 172–73) and Hacking (1983: 6); Hesse's list is not intended to characterize Kuhn specifically.
13. Musgrave 1980: 51.

For Kuhn$_1$ does not present a revised account of science as the construction and appraisal of theoretical representations. Kuhn$_1$ challenges this more general framework for dealing with science that Kuhn$_2$, his logical empiricist predecessors, and many of his post-empiricist critics share. Science is not primarily a way of representing and observing the world, but rather a way (or ways) of manipulating and intervening in it. Scientists are practitioners rather than observers. It is easy, of course, to overdraw this distinction. Scientists do construct theories and record observations. One can even argue that what is distinctive about scientific practice is the way it uses theories. As we shall see in chapter 4, however, some important differences in the understanding of what theories are result from emphasizing science as practice rather than representation. There is a difference between theoretically depicting the world and knowing your way about in it scientifically (in part by using theories).

To illustrate these changes, consider the following parallel summary of Kuhn$_1$'s most important innovations as a philosopher of science:

1. Scientific research presupposes an understanding of what one is dealing with, which cannot be made fully explicit in the form of theoretical representations; this understanding is embodied in scientists' skills, techniques, and projects. This general grasp is compatible with a variety of specific theories or even an absence of any specific theory in some cases. As both Laudan and Hacking have noted, even if we regard paradigms as theories, they are a different *sort* of theory from the specific, detailed hypotheses that scientists use to develop specific predictions or explanations.[14]

2. Observation in the philosophical sense of registering what one sees is not central to science. Being attentive to what is going on in the context of one's activities *is* important, but this attentiveness is influenced as much by one's practical concerns and craft skills as by one's theories. What is being criticized here is not an overemphasis on vision, but rather the philosophical account of vision as "observation." Scientific observation is much more like what Heidegger has called "circumspection" (*Umsicht*) than like reporting what one sees, however much one acknowledges the theory ladenness of the reports.[15]

3. The background knowledge scientists have established as reliable for and relevant to their research is not cumulative; many things

14. Laudan 1977: 71–72; Hacking 1983: 210–19.
15. Heidegger 1957: 69, 1962: 98–99; see also Hacking 1983: 167–69.

scientists once thought they knew are now dismissed as unintelligible, imprecise, irrelevant, unimportant, or incorrect.

4. There may be a narrow sense of "theory" in which the language of scientific theories may be precise; but in this sense theories constitute only a small part of scientists' understanding of the world. Scientists' grasp of paradigms gives them an understanding of the things they work with that is embodied in flexible but not fully articulable skills for dealing with them. The language they employ in exercising those skills is embedded in its use and its reference to concrete situations.

5. The point of scientists' activities and the significance of what they say depend upon how these are placed in a field of objects, techniques, instruments, skills, and concepts, a practical grasp of which is assumed in the everyday practice of research.

6. Scientists do not normally test theories; they use them. Such use is simultaneously the (increasingly controlled and precise) disclosure of phenomena and the refinement of the theories one employs. Specific claims are often disputed and tested, but this process presupposes substantial background information and technical judgment that is not subjected to test.

7. Discovery and justification are thus not separate activities, comparable to thinking up new ideas and then testing them in the laboratory. Discovery and justification are interrelated aspects of one activity, research, whose aim is the reliable disclosure and manipulation of its objects.

The shift from $Kuhn_2$ to $Kuhn_1$ is a transformation of the metaphors and models we use to understand science. $Kuhn_2$, and much of the philosophical literature from which it emerges as a reading of Kuhn's book, takes representing and observing to be the most characteristic activities of science. Scientists imaginatively construct models or theories that represent general features of objects in the world and how they behave (or of postulated objects that "save the phenomena" on instrumentalist interpretations). Using other models and theories as guides and a variety of physical apparatus as aids, they observe what happens in various situations (often ingeniously contrived), register this in appropriate inscriptions, and compare these inscriptions with what would be expected if their models and theories portrayed things accurately. The discrepancies that show up, and the new phenomena that appear, lead to the construction of new theories or models that are themselves tested in the same way, and so on.

Kuhn$_1$ replaces representing and observing with constructing, tinkering, and noticing as exemplars of scientific practice. Construction includes both the construction of phenomena or effects[16] and the construction of simulacra or models to guide our understanding of and intervention into those effects.[17] The two are closely intertwined, since the construction of phenomena is often an attempt to realize a physical counterpart to our models—to create an artificial and normalized situation within which tinkering with the model can parallel tinkering with the situation it models. Scientists tinker with instruments and experiments in order to obtain more consistent, reliable results; they tinker with their models in order to guide new or more precise experimental manipulations. In doing so, they must be attentive to what is going on around them. Being observant in this way is not a neutral recording of whatever they see but an *interested,* directed way of noticing what is relevant to their practical concerns. As Heidegger remarked, "'Practical' behavior is not 'atheoretical' in the sense of 'sightlessness'. The way it differs from theoretical behavior does not lie simply in the fact that in theoretical behavior one observes, while in practical behavior one *acts,* . . . [for] action has *its own* kind of sight."[18] Action also has its own kind of understanding, which cannot be reduced to theoretical representation. "Kuhn$_1$" as a reading of *The Structure of Scientific Revolutions* emphasizes the place of this practical understanding and practical sight in science. The rest of this book will attempt to explore some of the issues raised by such an emphasis on science as a field of practices. Such an approach to the philosophy of science situates it in a rather different philosophical tradition, drawing more extensively upon Continental European sources than is customary. It will bring to the fore some philosophically neglected features of science. Some traditional issues, such as realism, rationality, the structure and role of theories, and the continuities and discontinuities between the natural and human sciences, do remain central, but they receive somewhat different treatment than is perhaps customary. Above all, as I suggested in the introductory chapter, the place of scientific research and scientific knowledge in a political and social context will be moved to the center of philosophical reflection upon the sciences, from its current place somewhere beyond the margins.

16. Hacking 1983: 224–29.
17. Cartwright 1983: 143–62.
18. Heidegger 1957: 69, 1962: 99; see also Hacking 1983: 167–68.

Chapter 3

What Is Interpretation? Two Approaches to Universal Hermeneutics

HERMENEUTICS, or the theory of interpretation, has only lately emerged as an influence upon mainstream Anglo-American philosophy of science. Thomas Kuhn was one of the first to mention hermeneutics in this context, and even he noted that the word hermeneutic was not in his vocabulary as recently as 1972.[1] Much of this sudden upsurge of interest is due to the recognition that hermeneutics and pragmatism reinforce each other and even converge in some important ways. Thus it should not be surprising that the philosophers who have most extensively discussed the importance of hermeneutics for the philosophy of science—Rorty, Habermas, Bernstein, and Hesse—have also been prominently associated with the revival of pragmatism. The various versions of pragmatism that have emerged as responses to the collapse of empiricism can usefully be regarded as an attempt to universalize hermeneutics. This view is important, because pragmatism and the new empiricism, along with the parallel interpretations of Thomas Kuhn I outlined in the preceding chapter, can then be seen as alternative approaches to a universal hermeneutics. To see why the universalizing of hermeneutics is important and why the alternative approaches to science we have been discussing represent significantly different attempts at such universalization, we need to consider at least briefly the tradition in hermeneutics that would oppose any claim to its universality.

1. Kuhn 1977: xv.

Hermeneutics as the Epistemology of the Human Sciences

Hermeneutics has a long history, but it entered philosophical reflection upon the sciences only with the work of Wilhelm Dilthey.[2] Dilthey presented hermeneutics as a part of epistemology. He saw in it an alternative model of knowledge, one that could challenge the universality increasingly attributed to empiricist theories of knowledge modeled upon physics. This alternative model was of limited scope, however. Dilthey conceded the adequacy of empiricist accounts of the physical and biological sciences but insisted that extending empiricism to account for the scholarly investigation of human life and culture was illegitimate. His theory of interpretation was supposed to illustrate this point by exhibiting how a different model of knowledge could achieve objective knowledge of human beings, which also revealed aspects of human life that were inaccessible to empiricist inquiry. The prototype for his alternative model of knowledge was the interpretation of texts. We interpret texts on the assumption that they are already meaningful and that the task of interpretation is to discover and elucidate their meaning. Dilthey argued that many nontextual features of human life, such as actions, tools, social roles, and individual lives, can and should be taken as meaningful in the same way as texts are. Physics since Galileo had succeeded, according to the increasingly influential empiricist accounts of it, precisely by ignoring any meaning its objects have had for human beings. But in history and the generalized human sciences, these meanings are precisely what we need to understand. To set aside meaning would be to abandon one's very object of study. Sciences modeled on physics could describe a book as a physical object but not as a meaningful text, an action as a series of motions but not as a meaningful response to a situation, a life as a physical process but not as a unified life story. Dilthey thought that only by taking meaning seriously could we have any hope of understanding human beings and the social milieu in which they—we—live.

What model of knowledge is provided by the interpretation of texts? It cannot be a knowledge without presupposition. We first approach a text with the presumption that it is meaningful, that it says something, although we do not yet know what. We also bring to the text some understanding of what it is to speak in a language, and usually some grasp of the particular language in which the text is written. Our reading of the text will in turn appeal to our prior understanding of, for example, human beings—their character, the ways they interact, the kinds of things they say, what it makes sense for them to say, and so

2. Dilthey 1956, 1957.

forth. We bring the experience of a lifetime to our reading. When we begin to read, parts of the text will seem clearer than others. The clearer passages may then help us illuminate the obscure: What could someone who says this mean by that? As we read on, earlier passages are revealed in new ways. On occasion our interpretation of the opening of the text becomes untenable and must be revised because of what follows. On other occasions, we will see something in the beginning that originally escaped us. But that initial interpretation has provided the context for our reading of what followed; revise it and we transform our reading of the subsequent passages. Dilthey's version of the "hermeneutical circle" is at work here. We read each sentence, each word, with a presumptive sense of the whole text. But this sense of the whole can be developed only through our reading of individual words and sentences. Each continually transforms the other. A similar circle is at work between the text and the presumptions we bring to it. We already have some grasp of the language spoken and the things spoken about. This knowledge informs our reading but is in turn transformed by it. The text changes us, and this transformation in turn changes what we see in the text.

Once an interpretation is settled upon, what would justify a claim to knowledge of the sense of the text? We think we understand a text when our interpretation accords with our prior belief that the text is a meaningful whole. The text then makes sense. It is no longer confusing, obscure, or fragmented. The sense it makes, however, depends not just upon what is in the text, but also upon the presumptions we brought to it. This interpretation may make sense only to someone like me, who approaches it with presuppositions similar to mine and finds its emphasis in the same places. Charles Taylor notes the difficulty this limitation leads to.

> What if someone does not "see" the adequacy of our interpretation, does not accept our reading? We try to show him how it makes sense of the original non- or partial sense. But for him to follow us he must read the original language as we do, he must recognize these expressions as puzzling in a certain way, and hence be looking for a solution to our problem. If he does not, what can we do? The answer, it would seem, can only be more of the same. We have to show him through the reading of other expressions why this expression must be read in the way we propose. But success here requires that he follow us in these other readings, and so on, it would seem, potentially forever. We cannot escape an ultimate appeal to a common understanding of the expressions, of the "language" involved.[3]

3. Taylor 1979: 28.

There can thus be no finality to interpretation, nor any guarantee that agreement will be reached. But this lack of certainty does not mean that interpretation is merely private or subjective. The interpretation of a text concerns the sense a publicly accessible object has within the context of the public, social world. There is no certainty that our arguments for an interpretation will be conclusive, but there is the basis for an appeal to argument. As Dilthey argued, the main task of an epistemological theory of interpretation is "to counteract the constant irruption of romantic whim and sceptical subjectivity into the realm of history by laying the historical foundation of valid interpretation."[4] And this validity is rooted in the public accessibility of the text as the repository of meaning.

Many of the objects of interpretation in the human sciences are not themselves texts, of course. But actions, artifacts, social relations, and individual lives are analogues of texts in an important respect.[5] The terms in which we understand them, as clear or confused, significant or insignificant, are the same ones that guide our interpretation of texts. These various components of human life have a sense that can be expressed in words, even when not originally articulated this way. We interpret an action or artifact by saying what it means. This description proceeds with the same circular structure of presupposition and interpretation that characterizes the reading of a text. We interpret actions by using words; we interpret texts by using words different from the original ones. In either case, we understand them as already meaningful, and we take that same meaning to be expressible in a form different from the original.[6]

We can now enumerate four critical features of epistemological theories of interpretation, of which Dilthey's is one example. First, the domain in which interpretation functions is circumscribed by the notion of "meaning." Objects call for interpretation insofar as they have a meaning, which is not sufficiently clear but can attain clarity if further articulated. In Taylor's classic formulation, "The object of a science of interpretation must have a sense, distinguishable from its expression, which is for or by a subject."[7] This notion of "meaning by or for a subject" is notoriously problematic. Clearly more is meant than just that the concepts we employ in interpretation have an extension and an intension, which could be said of the concepts in any

4. Dilthey 1956: 337.
5. I owe this formulation to Taylor (1979: 25).
6. Taylor 1979: 26.
7. Taylor 1979: 27.

serious inquiry. It is the relation to human subjects that will be important here, and this relation will turn out to be difficult to characterize. Later in the book I will argue that it cannot bear the epistemological weight some neo-Diltheyans want it to shoulder. At this point, however, we can simply note that the domain of the "meaningful" in this richer sense has generally been taken to be coextensive with the actions, interactions, and productions of human beings.

The second characteristic feature of epistemological theories of interpretation has been their concern for questions of validity. How is it possible to achieve justified agreement about interpretations? What makes for a valid interpretation? How can idiosyncrasy in interpretation be limited or eliminated? What are the obstacles to justified consensus in interpretation, and how might they be surmounted? These questions, in short, reiterate the placement of interpretation theory within the theory of knowledge.

Third, the interpretation of meaningful objects is thought always to rest on some prior understanding of the social context within which their meaning functions. Dilthey characterized this knowledge as an understanding of life; we might find it more familiar if described as an understanding of a particular society and culture, and through this a grasp of what it is to be in a society or culture. I shall henceforth try to reserve the term "understanding" for this prior background against which interpretation takes place and "interpretation" for the activity of explication, which takes place against this background, and for the explication that results from this activity.

The fourth and final point is that, both in their constitutive notion of "meaning" and in their treatment of questions of validity, epistemological theories of interpretation have been posed in self-conscious opposition to empiricist accounts of scientific knowledge. Thus the conclusions of such theories can be usefully summarized in a series of contrasts to natural scientific knowledge as an empiricist might describe it. Consider the following partial list of alleged differences between the (empiricist) natural sciences and the (interpretive) human sciences.[8]

1. The natural sciences work with data that either are given identically to any observer or, in more sophisticated accounts, have conventional interpretations not normally in dispute. In each case they are objectively determinable such that any disagreement could be re-

8. A similar, shorter list was compiled by Hesse (1980: 170).

solved by empirical test. The human sciences can have no such indisputable data. They deal with meaningful objects and situations, whose interpretation is always potentially open to challenge based upon different interpreters' interests, situations, or prior beliefs.

2. Theories in the natural sciences are explanatory constructs from which observation statements are deducible. The human sciences produce not explanatory theories but redescriptions of the data, which attempt to reproduce their meaning in a clearer, more coherent way. They strive for understanding, not explanation.

3. The language employed by such theories in the natural sciences could in principle be formulated as a formal, uninterpreted calculus whose elements could then acquire univocal meaning by stipulation. The human sciences cannot avoid using the ordinary language within which its objects are constituted, with all its ambiguities and connotations.

4. Natural scientific theories are either confirmed or falsified by data that can be identified and described without reference to theory. The only significant relation between theories and data is that of subsumption. In the human sciences, "data" are manifested only within some interpretation, which they may nevertheless resist, compelling a new interpretation. The new interpretation may then reveal the "data" differently, and so on. The natural sciences thus leave their data as they are, while the human sciences continually reinterpret and remake theirs.

5. The natural sciences aim to eliminate anthropocentric references or connotations from their concepts, whereas the human sciences inevitably refer to human beings' interests, goals, beliefs, and feelings.[9]

6. The objects of the natural sciences are significant only as instances of a natural kind (or general type). The human sciences do not avoid generality, but they also must be concerned with the particular event, the individual life, or the local situation or culture. The one aims for universal knowledge, the other also for local knowledge.

7. The natural scientist is in principle anonymous. Who makes a claim is irrelevant to its validity or justification. But the human sciences always bear marks of their authorship, however implicitly. The experiences, cultural background, interest, and purposes of the scientist guide and help justify the interpretation. The natural sciences

9. Taylor (1980) makes this the central issue distinguishing the natural and human sciences.

thus mark a break with the concerns and practices of everyday life; interpretive human science is continuous with them.

It has long been disputed whether interpretation as described above is essential to the scientific study of human beings. More recently, however, defenders of interpretation in the human sciences have been challenged from a new direction, most forcefully by Rorty and Hesse. Most philosophers today would recognize, as Hesse has emphatically pointed out,[10] that these alleged distinctions between the natural and the human sciences presuppose a now largely discredited account of the natural sciences. Logical empiricism and its predecessors once laid claim to sovereignty over the legitimacy of any claim to knowledge. Now, however, they seem fundamentally inadequate even to account for the natural sciences, which they were originally developed to describe. Post-empiricist philosophies of science in many respects bear more resemblance to the epistemology of interpretive human science than to their own empiricist predecessors. The irony of this development has been aptly expressed by Taylor: "Old-guard Diltheyans, their shoulders hunched from years-long resistance against the encroaching pressure of positivist natural science, suddenly pitch forward on their faces as all opposition ceases to the reign of universal hermeneutics."[11] What Taylor does not recognize is that the claims of universal hermeneutics take two distinct forms.[12] More familiar to analytic philosophers is an account of interpretation as like translation, an account rooted in the work of Quine and Davidson on the one hand and Kuhn and Feyerabend on the other. It is this model of hermeneutics that is the background for the claims of Hesse and Rorty I cited earlier. All knowledge, including experiential knowledge, must be expressed in a language embodying theoretical presuppositions. There is no nonlinguistic, pretheoretical fact of the matter to which we could appeal to resolve disagreements about how the world is. Truth is a metalinguistic predicate. Deciding what is true is equivalent to deciding which sentences to accept. But sentences are acceptable only on the basis of their relation to other sentences we already accept. It is impossible for us to disagree radically with others about what is true, because a shared sense of what is the case is

10. Hesse 1980: 169–73.

11. Taylor 1980: 26.

12. This distinction was originally made by Dreyfus (1980), although he draws rather different conclusions from it than I will.

prerequisite to understanding what others say.[13] Truth and meaning thus become covariant concepts; I preserve or lose the truth of others' utterances by the way I render them in my language. And since there is no language-independent access either to what is the case or to what others mean, I can only do my best to bring the two together as consistently as I can. The problem is not to determine who is correct but rather to understand just what each side is saying. This is a problem of translation. The subsequent difficulty, determining which version to accept and use, will be partly resolved by translation. Any further resolution will require grounds other than what the facts (independently) are. It is this kind of view that Kuhn now claims was the basis for his account of scientific revolutions.

> Proponents of different theories are, I have claimed, like native speakers of different languages. Communication between them goes on by translation, and it raises all translation's familiar difficulties. . . . Without pursuing the matter further, I simply assert the existence of significant limits to what the proponents of different theories can communicate to one another. The same limits make it difficult, or, more likely, impossible for an individual to hold both theories in mind together and compare them point by point with each other and with nature.[14]

The other version of a universal hermeneutics has its origins in Heidegger and the later Wittgenstein. Interpretation is taken to be the working out of the possibilities open within a situation, rather than the translation of theories or beliefs. Hermeneutics is concerned with how one lives, and how one makes sense of how others live, rather than with the translation of beliefs from one vernacular into another. Where Quine, for example, takes the behavior of the "natives" in a situation of radical translation as evidence for or against various interpretations of their beliefs, for Heidegger it is what they are up to that calls for interpretation. What beliefs they may hold, or whether or not they hold beliefs or anything like a belief, is secondary. For that matter, understanding what someone is up to is more than just formulating an adequate description of their behavior that would let us predict it. Interpretation is a matter of coming to see what is at issue in how someone lives. In both cases, of course, the primary locus of interpretation is one's own case. For a post-Quinean, it is a question of my deciding which sentences to accept. For Heidegger, it is how I can be freed to encounter what is at stake, what is truly questionable, in living

13. Davidson 1984, essay 13; Rorty 1982: 3–18.
14. Kuhn 1977: 338.

now. The one universal hermeneutics interprets what is the case, the other what is the matter.

The recognition of this difference has also led to several recent attempts to resurrect the Diltheyan distinction between two kinds of inquiry. Hubert Dreyfus's position can most easily be described in terms of the two versions of universal hermeneutics I sketched above.[15] For Dreyfus, the hermeneutics of translation (theoretical holism, in his terms) provides a perfectly good account of how we come to understand the workings of the natural world. It may matter to us how the natural world is, but the world remains supremely indifferent to how it matters to us. Our concerns and practices and language affect the kind of account we give of the natural world, but we do not give an account of it as affected (in its general structure, at least) by our concerns, practices, or language. Things are otherwise with the study of human beings. Here we are concerned precisely to capture what is at issue in our everyday practices. A description that rendered with perfect accuracy what we do—in the sense of describing our movements, for example, or even in describing what goes on in our institutions and practices while taking their point for granted—is not satisfactory. What we need to get clear about is not just what we do, but the point of doing it, and how such a point takes hold over us and gives meaning to what we do. The difference discerned between the study of human beings and the sciences of nature is not that one is interpretive and the other not. The difference is supposed to lie in how they are interpretive and what the point of the interpretation is. The interpretation of human beings matters to us in a way the interpretation of nature does not, and this importance fundamentally changes the character of the interpretation.

Charles Taylor attempts to draw a similar distinction by arguing that the study of human beings is an interpretation of something (the human social world) that is self-interpreting and that is therefore concerned with the outcome of the interpretation.[16] Since this concern must be reflected in the interpretation if the latter is to be comprehensive, the interpretation of human beings must be different in kind from the interpretation of natural events. Like Dreyfus, Taylor attempts to work out in some detail the consequences of this view for the study of human beings. Jürgen Habermas, in many ways crucially at odds with Taylor and Dreyfus, nevertheless develops a parallel point here. There is a fundamental difference between knowledge of human

15. Dreyfus 1980, 1984.
16. Taylor 1979, 1980.

beings and knowledge of the natural world, because they satisfy different cognitive interests.[17] We study human beings to satisfy our communicative interest in understanding other persons, with the aim of achieving "unconstrained agreement and non-violent recognition."[18] We study other objects to achieve technical control over them for our own ends. In the one case knowledge aims to determine through conversation what our ends are. In the other case its aim is to achieve those ends through manipulating and controlling the world around us.

The rest of this chapter will be devoted to a more detailed contrast of the two ways of universalizing hermeneutics, which will provide the basis for much of what follows. Detailed discussion of Dreyfus, Taylor, and Habermas must be postponed until chapter 6, by which point we will have seen how these alternative hermeneutics affect our understanding of the natural sciences. In the two intervening chapters I will argue that a Heideggerian hermeneutics of practice can provocatively appropriate many insights of the new empiricism. This approach will have important consequences for understanding the attempts to distinguish two kinds of inquiry in the natural sciences and the human studies, as well as for the other central issues raised by post-empiricist philosophy of science. Above all, it will focus our understanding of the sciences upon what I take to be its most fundamental issue, the relation between knowledge and power or, more broadly, between epistemology and political thought.

Hermeneutics as Universal Theory

In this section I shall attempt to show how contemporary pragmatism stemming from recent philosophy of science can be seen to extend hermeneutics beyond the specialized domain of the human studies. The position as I will present it owes much to the work of W. V. O. Quine. Quine's position is not yet thoroughly hermeneutical, however, because of the confidence he retains in the relative stability of observation statements.[19] Thus Rorty's account of universal hermeneutics unites Quine's epistemological holism with Sellars's rejection of the myth of the given.[20] Hesse, whom I shall cite more often, adopts a generally Quinean position but accepts that observation statements are more theory laden than Quine would be inclined to admit.[21] It may

17. Habermas 1968a, 1968b, 1970, 1971.
18. Habermas 1968a: 222; 1971: 176.
19. Quine 1975: 88–90.
20. Rorty 1979, esp. chap. 4.
21. Hesse 1980. Hesse also breaks with this position to some extent with her account of the pragmatic criterion.

seem strange to many philosophers to find these views, especially Quine's, examined under the label hermeneutics. But the parallel has already been suggested by Hesse, Rorty, and Bernstein, and the reasons for it should emerge clearly in the ensuing discussion.

Traditional accounts of hermeneutics as the epistemology of a particular region of knowledge (the *Geisteswissenschaften*) distinguish sharply between the artificial language of the natural sciences and the ordinary language of human interaction, whose sense is bound to our understanding of life as participants.[22] Empiricist theories of science in turn came to describe the language of science as itself twofold. Theories were expressed in formal languages, for which extratheoretical correspondence rules stipulated the meaning of its terms; scientific observation was described in its own "language," whose predicates could be unproblematically applied to empirical situations by any competent observer and speaker of the language.[23] Ordinary language contained the observation language of science, but the Diltheyan hermeneuticists considered this sublanguage insufficient for the purposes of the human sciences.[24] Theoretical hermeneutics collapses both of these distinctions by insisting that everyday knowledge and scientific knowledge are not different in kind. Quine, for example, proclaimed:

> We imbibe an archaic natural philosophy with our mother's milk. In the fullness of time, . . . we become clearer on things. But the process is one of growth and gradual change: we do not break with the past, nor do we attain to standards of evidence and reality different in kind from the vague standards of children and laymen. Science is not a substitute for common sense, but an extension of it.[25]

It would be more accurate to his views, however, to say that common sense contains a protoscientific theory. For Quine's claim is that both scientific understanding and the understanding of everyday life are embodied in statable hypotheses about how the world is. "Hypotheses in various fields of inquiry may tend to receive their confirmation from different kinds of investigation, but this should in no way conflict with our seeing them all as hypotheses. We talk of framing hypotheses. Actually, we inherit the main ones, growing up as we do in a

22. Dilthey 1957: 82–83, 90, 118.
23. Suppe 1977: 66–86.
24. Taylor 1979: 32–55.
25. Quine 1976: 229.

given culture."[26] Not all our hypotheses are explicitly formulated, but they are inscribed in our behavior and our dispositions to behave, and if need be they could be explicitly formulated. Their common character as hypotheses is unaffected by whether they have been made explicit.

Interpretation, on this account, thus consists of forming a hypothesis. It does not matter whether the hypothesis concerns the motion of particles in a gas, the action of an individual, or what is said in a text.[27] Even observation statements are to be construed as hypotheses. They do not wear their meanings on their sleeves but must be interpreted against the background of other hypotheses. Observations enter into the realm of knowledge only when they have been formulated in publicly accessible terms—when they have been given hypothetical formulation in a language. There is still a distinction between observation and theory, but it is only a relative one. Observation statements are those hypotheses that are formulated more directly as responses to particular sensory stimulations. But what counts as a direct result of sensory stimulation is itself subject to theoretical interpretation.[28]

A basic postulate of empiricism is still preserved within theoretical hermeneutics: "Whatever evidence there is for science is sensory evidence."[29] And what is true for science is true for all our other hypotheses and beliefs. But this evidential relation is considerably complicated. To begin with, hypotheses do not have their own empirical evidence that by itself determines their acceptability. Quine expressed this point in two famous metaphors:

> The totality of our so-called knowledge or beliefs, from the most casual matters of geography and history to the profoundest laws of atomic physics or even of pure mathematics and logic, is a man-made fabric which impinges on experience only along the edges. Or, to change the figure, total science is like a field of force whose boundary conditions are experience. A conflict with experience at the periphery occasions readjustment in the interior of the field. Truth values have to be redistributed over some of our statements. . . . [But] no particular experiences are linked with any particular statements in the interior of the field, except indirectly through considerations of equilibrium affecting the field as a whole.[30]

The well-known reason for this state of affairs is the complexity of any actual empirical test situation. A theoretical hypothesis cannot be

26. Quine and Ullian 1970: 81.
27. Follesdal 1979.
28. Hesse 1974: 20.
29. Quine 1975: 75.
30. Quine 1953: 42–43.

used to predict the outcome of a test without presupposing some statement of the initial test conditions. The determination of these conditions in turn relies upon some theoretically supported lawlike regularities, evidence for which also depends on a growing set of theoretical assumptions. Without appeal to such regularities, we could not rule out other interpretations of the outcome of the test as due to factors other than those proposed in the original hypothesis. If the outcome of the test then conflicts with the original hypothesis, the test itself cannot uniquely assign the conflict to a particular source. The original hypothesis may itself have been mistaken, but the conflict might instead be traceable to an auxiliary hypothesis or to one of the theories that provided part of its justification. Thus Quine concludes: "And how wide is a theory? No part of science is quite isolated from the rest. . . . Legalistically, one could claim that evidence counts always for or against the total system, however loose-knit, of science. Evidence against the system is not evidence against any one sentence rather than another, but can be acted on rather by any of the various adjustments."[31]

This situation becomes still more complicated when one recognizes the influence of theory upon our interpretation of the observed evidence. The evidence itself can be stated only in a language whose application presupposes the acceptance of various lawlike regularities. A conflict between hypothesis and test outcome can thus not just be deflected onto other background hypotheses, but be taken as evidence against our description of the test outcome. Hesse concludes that

> no feature in the total landscape of functioning of a descriptive predicate is exempt from any modification under pressure from its surroundings. That any empirical law may be abandoned in the face of counterexamples is trite, but it becomes less trite when the functioning of every predicate is found to depend essentially on some laws or other and when it is also the case that any 'correct' situation of application—*even that in terms of which the term was originally introduced*—may become incorrect in order to preserve a system of laws and other applications.[32]

Once we accept the theory ladenness of observation in this strong form, those statements that are relatively observational and those that are relatively theoretical function together as a single network of hypotheses.

Hesse has suggested that such a network could be modeled by a

31. Quine 1970: 5.
32. Hesse 1974: 16.

self-reprogramming learning machine.[33] The machine is a model of how a new, more coherent representation of the world is developed on the basis of prior representations and new empirical input. The model maps representations onto representations and hence is indifferent to who is supposedly doing the representing, that is, "whether that device is conceived as the individual scientist, the whole body of scientists, or the institution of science represented by its learned societies, journals, and textbooks."[34] The machine has the following essential features:

1. Physical interaction with its environment.
2. A program for coding into machine language the way the environment affects it.
3. Coherence conditions that specify the desiderata for more complex classifications of its coded environmental input. Hesse said that "the function of the coherence conditions is to produce from the initial classification a 'best theory', or range of best theories, conforming optimally both to the initial classification and to these conditions, where 'optimally' is itself defined by the conditions."[35]
4. A stock of previously coded (and further classified) input.
5. An "external feedback loop" that compares the predictive output of its more complex classifications with its empirical input.
6. "Internal feedback loops" that adjust the machine's program to deal with mismatches between predictive output and empirical input.[36]

Hesse allows for three kinds of internal feedback loop. The simplest kind reinvestigates anomalous input to attempt to reduce mismatches through marginal readjustment of the programming of input.[37] A more complicated feedback loop would permit more substantive changes in the programming of input, comparable to redefining the predicates of the classification language. In such a machine the coherence conditions would govern changes in the observation language in order to satisfy better their specification of what counts as good theory.[38] Third, the machine may have a final loop that, in extreme cases of mismatch, could adjust the coherence conditions themselves

33. Hesse 1980: 125–28.
34. Hesse 1980: 125.
35. Hesse 1980: 126.
36. The basic model is presented in Hesse (1980: 125–27).
37. Hesse 1980: 127.
38. Hesse 1980: 127–28.

"in the light of success and failure of the sequence of best theories in accounting for the available observation sentences, and in making successful predictions."[39]

What I am calling "theoretical hermeneutics" is a view of interpretation as modelable by such a learning machine of the third kind, that is, one possessing all three kinds of internal feedback loop. Theoretical hermeneutics takes interpretation to be the coding or reclassification of how the world impinges upon the interpreter. Such a hermeneutics is a theory about how we acquire our best theoretical representation of the world, and it is itself part of that best theory. Thus, as Quine insists, epistemology is not "first philosophy" but constitutes part of our scientific theorizing.[40] The goal of interpretation according to this model is truth, the accurate representation of what is the case. But as Hilary Putnam has forcefully argued, the notion of "truth" has no epistemological force apart from the criteria we employ in deciding when we have achieved it,[41] that is, apart from what Hesse has called "coherence conditions." And these in turn are part of our total theory.

The import of this position can be seen in terms of Quine's criticism of "the idea idea." Quine is arguing that when we describe how the world impinges upon us, or when we provide a theoretical account of some portion of the world, there are no "meanings" for our interpretation to correspond to. There is no prelinguistic fact of the matter by which our interpretations can be judged and no objects to which our terms refer independent of a background of language and theory.

What makes sense is not to say what the objects of a theory are, absolutely speaking, but to explain how one theory of objects is interpretable or reinterpretable in another.[42] It is not just that we cannot ascertain for certain what others are talking about and can only offer another reading of it in our own terms. The same is true in our own case. "But if there is really no fact of the matter, then the inscrutability of reference can be brought even closer to home than the neighbor's case; we can apply it to ourselves. . . . After all, as Dewey stressed, there is no private language."[43] Quine's position can thus be equally well characterized in terms of Taylor's formulation of the hermeneutic circle: "What we are trying to establish is a certain read-

39. Hesse 1980: 128.
40. Quine 1975: 82–83.
41. Putnam 1981: 129–30.
42. Quine 1975: 50.
43. Quine 1975: 47.

ing [of how the world is], and what we appeal to as our grounds for this reading can only be other readings."[44] There can be no successful "attempt to reconstruct knowledge in such a way that there is no need to make final appeal to readings or judgments which cannot be checked further."[45]

Theoretical hermeneutics, like Diltheyan theories, is an epistemological theory. But it is not an account of knowledge within a special domain, such as that of texts and text analogues. It provides a general account of knowledge (albeit one that is on its own admission corrigible if a better theory of the world results). Texts hold no special place for it as objects of interpretation. The indeterminacy of translation is equally a problem for our reading of Shakespeare and our reading of bubble-chamber photographs, or even of the color of an object in front of us. For Quine, admittedly, there is a difference between the indeterminacy of translation and the underdetermination of physical theory by evidence, because he insists that "theory in physics is an ultimate parameter . . . so we go on reasoning and affirming as best we can within our ever under-determined and evolving theory of nature, the best one that we can muster at any one time; and it is usually redundant to cite the theory as parameter of our assertions, since no higher standard offers."[46] Quine suggests that matters are different with understanding what other people mean, because even if we settle upon our theory of nature, translation would still be indeterminate. "Where indeterminacy of translation applies, there is no real question of right choice; there is no fact of the matter even to *within* the acknowledged under-determination of a theory of nature."[47] But it is not clear what kind of difference this indeterminacy makes. There is also no higher standard for *translation* than the best hypothesis we can muster at any one time. And unless one simply assumes (as Quine does) that our best theory of nature takes epistemological priority over our best translations, we could just as well point out that our theory of nature would still be indeterminate within the limits of our translations. Quine gives priority to physics because of its relation to a relatively stable stock of observation sentences; but once one takes seriously the revisability of our observation vocabulary, then interpretations of nature and interpretations of utterances are in the same situation. Both acquire whatever determinacy they have against a taken-for-granted background of theory. We may well be more confi-

44. Taylor 1979: 28.
45. Taylor 1979: 29.
46. Quine 1969: 303.
47. Quine 1969: 303.

dent in our interpretations of nature and may well have good reason for our confidence. But this assurance has much to do with the particular successes and failures of our best current theory and nothing to do with deep differences in the kind of interpretation required.[48]

We can now sum up the salient features of this position as a theory of interpretation and its implications for a Diltheyan distinction between the epistemologies of the natural and the human sciences. Theoretical hermeneutics is an account of how we interpret things, behavior, and utterances by situating them within a previously understood context. New interpretations (often prompted by new experiences) gradually shift this prior understanding, which in turn allows things to be interpreted in new ways. Six central questions must be asked about any such account: (1) What is this previously understood context within which all interpretation is situated? (2) What kind of "understanding" do we have of it? (3) "Who" is it that understands? (4) What does interpretation consist in? (5) What are they interpretations of? (6) What is at stake in these interpretations? Theoretical hermeneutics takes the context to be a network of beliefs and values (e.g., the program and memory of Hesse's learning machine), expressible in a language although in practice never fully articulated. Understanding takes the form of behavioral dispositions to use sentences in particular ways, to respond in particular ways to various situations, to use some terms rather than others in describing those situations, and so forth. We take up such dispositions as competent speakers of a language (or languages), who have inherited the beliefs of a culture. That is, we speak as individuals but can do so only intersubjectively; a shared language and background of beliefs and values is necessary for us to be intelligible to others and to ourselves. There is no such thing as a private language, for though beliefs are held by individuals, the criteria for what it is to hold a belief and whether one actually holds any particular belief are social. Understanding is thus equivalent to the *acceptance* of a background of beliefs and values, where "acceptance" includes both the ability to produce or employ them on (socially) appropriate occasions and the disposition to do so. Interpretation consists in formulating (or behaving in accord with) new beliefs or reformulating old ones. As Quine might put it, it is to formulate or act upon a hypothesis. What we interpret is the world, taken to be a universe of objects and events that exists independent of what we do and say, but that is manifest to us only through our previously understood beliefs and values. What is at stake in our

48. A similar argument is developed in Rorty (1979: 202–3).

interpretations is the truth of our descriptions of this world. But since truth cannot be assessed independent of the coherence conditions or values internal to our network of prior belief, what is at stake ultimately are the values posited within our interpretations themselves.

On such an account there is no epistemological difference between interpretations of nature and interpretations of human beings and their institutions and practices. Both belong in the same way to our theoretical network of beliefs about the world. Both contain hypotheses that acquire their sense, and their truth values, from social practices of language use, behavior, and evaluating beliefs. There may be important differences between the natural world and the social world as we understand them; but these would be empirical differences that are established by our theories, not some kind of transcendental difference between two different kinds of knowledge about two different kinds of objects. Dilthey's distinction between the *Geistes-* and *Naturwissenschaften* thus collapses within a theoretical hermeneutics.

Heidegger's Hermeneutics of Practice

Heidegger describes two fundamentally different senses of "hermeneutics" in *Being and Time.* In division 1 of this book, he describes human existence as itself hermeneutical. Our way of being-in-the-world embodies an interpretation of the world and of ourselves, which can itself be elucidated by the interpretation of our everyday practices. Thus both the attempt to disclose the meaning of our practices and the practices themselves are hermeneutical. Heidegger thereby "emphasized that hermeneutics . . . does not have its usual meaning, methodology of interpretation, but means the interpretation itself."[49] In division 2, he explores a "deeper" version of hermeneutics. This deeper interpretation reflects the apparent discovery that our everyday interpretations, which are what we make of ourselves and the world, reflect an attempt to disguise the "uncanniness" or lack of grounding of those interpretations. Hermeneutics in this latter sense represents the unmasking of this disguise, permitting an "authentically resolute" existence, which faces up to the uncanniness of being-in-the-world rather than fleeing from it. In his later works Heidegger still accepts the first account of everyday activity as hermeneutical, but he abandons the second sense of hermeneutics as an unmasking of a hidden truth. The uncanniness he once thought was an essential characteristic of being human he later interprets as symp-

49. Heidegger 1959, 1971: 28.

tomatic of the rootlessness of the interpretations embodied in the practices of modernity. In what follows, we will be concerned only with the first sense of "hermeneutics," in which our everyday activities are an interpretation grounded in a prior understanding of what it means to be. For it will be in this sense that scientific practices can be regarded as hermeneutical, in a way alternative to a Quinean theoretical hermeneutics. We shall reserve to a later chapter any discussion of this rootlessness that Heidegger takes as characteristic of what our modern practices are doing to us.

What do we mean when we say that our everyday practices embody an interpretation of the world? To begin with, we take account of things around us in a host of specific ways. We use equipment, and in this use it acquires an orientation, a focus, a significance, a function. We also avoid things, take note of them, care for them, discard them. Both what things are (clocks, lights, songs, flowers, corridors, etc.) and how they are (delicate or sturdy, objects of reverence or disdain, things that belong or things that are out of place, events striking or unnoticed) show up in the ways we deal with what surrounds us. We differentiate persons from things in various ways, taking account of them in our glances, gestures, posture, and stance as well as in our words and deliberate actions. The many subtle nuances of social station, situation, personality, and character show up in our most mundane ways of comporting ourselves. We interpret ourselves and the world in what we do and how we do it; our everyday practices and our bearing as we engage in them make us what we are and repeatedly remake us.

We likewise inhabit space and live time in quite specific ways. Our space is oriented by our vertical posture, our forward direction, the range of our grasp, and our mobility.[50] The specific spaces we inhabit are organized to accord with what we generally do within them. Things are near or far, in front or behind, in our way or to the side, large or small (relative to what we do with and around them). We sometimes crowd a room, moving expansively and using the entire space; we can also withdraw and allow the space to overwhelm us. Spaces can be threatening, comforting, alienating, energizing, or lifeless. Time can pass serenely or in a frenzy. It can be tightly organized or "free flowing," compressed or expansive. Our everyday ways of engaging the world exhibit a style, which is itself an interpretation of what it is to be and how things are in the world.

Our practices and the interpretations they embody hang together.

50. Heidegger 1957, secs. 22–24; see also Todes 1966.

Or rather, any particular activity acquires its interpretative sense and its intelligibility from the coherence of practices, roles, and equipment to which it belongs. This dependence is clearly there with equipment. As Heidegger remarked:

> Taken strictly, there 'is' no such thing as *an* equipment. . . . Equipment always is *in terms of* its belonging to other equipment. These 'things' never show themselves proximally as they are for themselves, so as to add up to a sum of *realia* and fill up a room. What we encounter as closest to us (though not thematically) is the room; and we encounter it not as something 'between four walls' in a geometrical spatial sense, but as equipment for residing. Out of this the 'arrangement' emerges, and it is in this that any 'individual' item of equipment shows itself.[51]

The "room" in this sense is not just another, more complicated thing whose components are pieces of equipment. The room is the space or field within which the equipment belongs. Heidegger is trying to articulate a sense of contextuality as determinative of what things are without treating contexts as themselves things.

What is true for our equipment is also true for the practices in which we use it and the roles we take up in doing so. Particular social roles presuppose the existence of related roles, of the equipment necessary to take up these roles, and of a shared sense on the part of the various participants of how to take them up. One cannot interpret oneself as a teacher unless there are also students (and perhaps also administrators, janitors, publishers, tuition payers, etc.). Or perhaps better, the role of a teacher takes shape in relation to coexisting roles and available equipment and institutions. The contexts of roles, practices, equipment, and goals within which our activities function both guide and make sense of what we do. Alasdair MacIntyre has attempted to articulate this contextuality of action in terms of shared "schemata" for interpretation.

> Consider what it is to share a culture. It is to share schemata which are at one and the same time constitutive of and normative for intelligible action by myself and are also means for my interpretations of the actions of others. My ability to understand what you are doing and my ability to act intelligibly (both to myself and to others) are one and the same ability. . . . [These schemata] are not, of course, empirical generalizations; they are prescriptions for interpretation.[52]

Heidegger, however, would hasten to add that these schemata are not normally *explicit* prescriptions or rules. They are a generalized, flexible

51. Heidegger 1957: 68–69, 1962: 97–98.
52. MacIntyre 1980: 54–55.

know-how, an ability to recognize "what one does" in a given situation and to do it, without necessarily being able to give a general account that would explain one's recognition in any particular case. Our understanding of the social "space" within which our interpretive activities are intelligible is not a set of beliefs or internalized rules, not something "cognitive" in the traditional sense, but an ensemble of skills inscribed in our bodies. These skills are ways of inhabiting a meaningfully delineated social world, whose exercise in turn renews and reinforces the delineations.[53]

We need to explore further this difference between skills and practices, on the one hand, and beliefs, dispositions, and rules on the other. It is well known that we possess and exercise many skills we cannot give a full account of. From riding a bicycle to producing a coherent and grammatical sentence, from recognizing a complex visual pattern to solving a physics problem, a multitude of our practices involve inarticulable skills. We can often suggest hints or maxims, but seeing the point of the hints and being able to follow the maxims to a large extent presuppose the skill they are supposed to account for. As Dreyfus has pointed out, once we actually have acquired a skill we do not use such maxims, and we often violate the very ones that helped us catch on to the skill in the first place.[54] We acquire skills not by learning and applying rules or by acting in accord with beliefs, but by imitation and habit. We are socialized into our skills, doing what we see those around us doing; we acquire facility and the ability to make the requisite discriminations in the course of acting them out.

This process, however, suggests incorrectly that skills are, as Plato thought, a blind knack or disposition. This suggestion belies the flexibility of skills. In learning to throw, one learns not a repetitive series of movements, but a range of responses to throwable things. Having learned to throw overhand, I can then easily throw three-quarters or sidearm, even though the motions required are quite different. Having learned to imitate a fairly limited set of sentences, I can then produce an unlimited variety of different ones. What one learns is not the repetition of an actual movement or thought pattern, but the grasp of a field of possibilities. The difference between holding an actual belief or following a rule you have before you and having a possibility "in hand" is crucial to Heidegger's hermeneutics. He begins the discussion of interpretation in *Being and Time* by saying: "As

53. For a detailed working out of this argument, see Dreyfus (1979, preface and part 3).

54. Dreyfus 1984: 6–7.

understanding, Dasein projects its being upon possibilities. This *being-towards-possibilities* which understands is itself a capability to be. . . . The projecting of the understanding has its own possibility—that of developing itself. This development of the understanding we call "interpretation," . . . the working out of possibilities projected in understanding."[55] Heidegger is not referring to the explicit posing and weighing of possibilities among which we then make deliberate choices. The projecting of the understanding is rather that "stance" toward the world and that corresponding "set" of the world *out of which* specific possibilities can emerge as up for choice and *upon which* we choose. We always find ourselves in a world whose sense is already laid out toward concrete possibilities. We choose among specific possibilities, but we do not in the same way choose the field of possibilities from among which we choose. This field remains hidden from us, not in the sense of something ineffable or mysterious, but as something so close to us and so obvious that we see right through it. We are unable to envisage concretely what an alternative to this field would be, and we are likewise unable to envisage the field itself as such.

At first consideration, theoretical hermeneutics may seem to make a similar point by recognizing that some of our most fundamental beliefs are inherited rather than chosen and are so basic to us that we would modify other beliefs in order to protect them from falsification. These fundamental beliefs are the quasi-analytic core of a conceptual scheme, the paradigmatic beliefs that supposedly form the basis for normal science and everyday life. But even if these beliefs are taken for granted, they are still beliefs like any others in being articulable and having a truth value. If they are always taken to be true, that is because of our commitment to them. The fundamental place held by these beliefs is the responsibility of the believer. Contrary to this position, Heidegger is claiming that our possibilities emerge not from fundamental beliefs, but from a way of being in the world. It is more like what Wittgenstein has called a "form of life" and Hacking a "style of reasoning."[56] A form of life is not true or false, nor is a style of reasoning. It is, as Hacking puts it, what determines what is true-or-false—a candidate for truth or falsity. Likewise we do not choose a form of life or style of reasoning. We inhabit it, and only from within it do possibilities emerge as up for choice. Thus the differences between theoretical and practical hermeneutics do not just represent alternative ways of picking up the same stick. It is not

55. Heidegger 1957: 148, 1962: 188–89.
56. Wittgenstein 1953, par. 241; Hacking 1982: 49–51.

just that Quine sees all of our behavior as embodying implicit beliefs while Heidegger sees belief as one of myriad ways of comporting ourselves. It is that Quine sees the whole of our beliefs as one total theory that is itself a complex belief while Heidegger sees the context of our behavior as what we cannot choose or act upon. There is no such *thing* as the total context of our behavior.

There is a serious danger, however, that we will mistake Heidegger's notion of "understanding" for something numinous and ineffable somehow lying "behind" the possibilities we actually have before us. It would then express some deep, mysterious truth that we could get at only indirectly. This misunderstanding, I believe, is responsible for much of the suspicion of Heidegger still prevalent among Anglo-American philosophers. If we must persist with the metaphor of surface and deep truths, then Heidegger is talking about something that is on the surface. We do not need to look "behind" our everyday practices for some hidden understanding. Heidegger is talking about what is at work in everything we say or do and can be made manifest in our everyday grasp of things and our ways of dealing with them. What he denies is that this understanding can be grasped as something formal or otherwise abstractable from our actual involvement with one another in the world. "Understanding" in Heidegger's sense is always local, existential knowledge. In calling understanding local and existential, I mean that it is bound to concrete situations, embodied in an actual tradition of interpretive practices carried on from generation to generation, and located in persons shaped by specific situations and traditions. Understanding is thus not a conceptualization of the world but a performative grasp of how to cope with it.

What is understood is the way one's actual situation hangs together and makes sense as a field of possibilities for interpretation. Above all, it is the way one's situation has a *direction* to it, pointing beyond itself toward future possibilities. (We do not usually regard them as possibilities for interpretation, of course, but rather see them as possibilities for doing this, that, or the other; only philosophers, who bring different concerns to this situation, will seize upon the commonality of these various doings and various thises and thats as ways of interpreting oneself and the world.) The German word *Zusammenhang* captures this point better than its usual English equivalent, "context." The English word suggests that there is something *additional* that must be comprehended, whereas what Heidegger is trying to point out is that it is the way things hang together as a meaningful situation within a form of life that allows the individual things to be identified and to make sense. The context within which we understand and interpret the world is not something *other* than the things within it, or even something(s) among them. It is the configuration and direction of the things, which is what enables them to be manifest as things in the first place. Another way of getting at this is to see that according to a theoretical hermeneutics, *some* basic beliefs and values must be presupposed in order for others to make sense and to show up as true or false, but *which* beliefs and values these are may be arbitary. We can in principle recognize them as theoretical presuppositions that we have chosen (however implicitly) and could abandon in favor of others. For Heidegger, by contrast, the configuration of the world (and hence the way things show up for us) is not something we have chosen and not something we can articulate. It is therefore not something we could "stand back from" and accept or reject. It is what provides us with a hold on the world, allowing us to make sense of ourselves and to encounter significant things around us. To stand back from it would be to lose our grip rather than to make our interpretation of things clear. It is not a set of beliefs or assumptions we have, but a way into the world that "has" us. Thus it is not accidental or arbitrary, not one "conceptual scheme" among others. This configuration of things is the manifestation (to us) of what it is to be. Such a configuration may change over time, but not as the result of deliberate choice or action.

The significance of this last point emerges by contrast to theoretical hermeneutics. Theoretical hermeneutics takes interpretation to be a concern for what is the case, reflected in the attempt to represent things accurately. Practical hermeneutics takes interpretation to be a concern for what matters, reflected in the attempt to live meaningful lives. Thus, in the former case, what matters comes to be understood as values. Dreyfus has concisely summarized Heidegger's challenge to this development:

> Values are objective, explicit options that we can stand outside of, picture, and choose among. . . . Once we get the idea that there is a plurality of values and that we choose which ones will have a claim on us, we are ripe for the modern idea, first found in Nietzsche, that we *posit* our values—that is, that valuing is something we do, and value is the result of what we do. Once we see that sets of values or mind sets or world pictures are simply posited they lose all authority for us and, far from giving meaning to our lives, they show us that our lives have no intrinsic meaning.[57]

In Hesse's learning-machine model of theoretical interpretation, for example, the coherence conditions or values governing the interpreta-

57. Dreyfus 1981: 511–12.

tion are simply one aspect of the interpretive program, changeable by feedback loops just like any other. They are part of the overall schema for depicting what is the case. As such, they do not matter to anyone at all. This situation is reflected in the model's indifference to who or what it is a model of; the "subject" who interprets is abstracted from any real involvement with what it interprets. In Heidegger's case our actual involvement with a situation that matters to us is what governs the interpretation. Interpretation is existence, the working out of what it is to be a person here and now. We come to consider what is the case only in the course of taking our stand upon what it is to be. It is not that we thereby determine what is the case, for we do not. It is rather that we discover what is only by how it affects us. Things matter to us, we are vulnerable to them, and only thereby can we make sense of what they are.

The same six questions I used to sum up my account of theoretical hermeneutics will help focus the contrast between it and the practical hermeneutics I have attributed to Heidegger. To begin with, the context within which interpretation takes place is a configuration of equipment, persons (who already occupy specific social roles), and physical setting rather than a "web of belief." This configuration is not just a collection of things, but a setting that already has a focus and a direction—that is already opened toward possibilities. Heidegger calls this setting a "world," where "world can be understood . . . not as those beings which Dasein essentially is not and which can be encountered within-the-world, but rather as that '*wherein*' a factical Dasein as such can be said to 'live'."[58] Unlike theoretical hermeneutics, which sets interpretation against a background of representations, Heidegger sets it within a configuration of presences and absences. Representations have a place within this configuration, but their place is neither privileged nor universal. Heidegger would insist that we do encounter things unmediated by theories or hypotheses, but not that we thereby encounter them without presuppositions. What we presuppose is a form of life.

In *Being and Time,* Heidegger suggests that understanding contributes a threefold structure to any interpretation, which he calls the interpretation's prepossession (*Vorhabe*), preview (*Vorsicht*), and preconception (*Vorgriff*).[59] Prepossession is our general prior familiarity with the things, practices, and roles available within the world. It provides our sense of what we have before us (it is also a grasp of the

58. Heidegger 1957: 65, 1962: 93.

59. Heidegger 1957: 150, 1962: 191. An interpretation that differs significantly from what follows can be found in Dreyfus (1980: 10).

world that "has" us, since, as Heidegger insists, we find ourselves thrown into this world that is not of our own making). Preview is our sense of how to proceed, of what possibilities might be open to us and how they might be taken up. As I have pointed out, it is not an articulation of definite possibilities we can choose among but a way of being in the world from which certain possibilities emerge as real ones. "This preview 'takes the first cut' out of what has been taken into our prepossession, *with a view to* [*auf*] a definite way in which this can be interpreted."[60] This preview then becomes definite with regard to our preconception of what would count as an adequate interpretation. Preconception is a more or less definite anticipation ("anticipation" is an alternative translation of *Vorgriff*) of the completed interpretation and of what would count as success or failure.

Understanding takes the form of a skillful knowing our way about in the world rather than theoretical knowledge of the world. The difference is subtle, but Heidegger takes it to be of utmost importance; it has to do with power. If our understanding consists of beliefs we can accept or reject and values we can choose or turn away from, we become masters of ourselves and our world. We are entirely the product of our choices, and the world is something standing over against us as object. "Values" become inner and experiential, and the world is valuable only insofar as we "give" it value. Hence the world becomes something manipulable for the sake of the values we have chosen. It becomes the stock of resources on hand for the fulfillment of what we value.[61] Heidegger, by contrast, sees our understanding as situated within a field of possibilities that we have not chosen and that are not up for choice. Who we are is not our own doing, but what we do and what happens to us nevertheless matters to us. The world is the setting within which we encounter meaningful possibilities; but it is the world that is meaningful and not just our inner experience or values.

As interpreters, Heidegger would characterize us as embodied, vulnerable persons, shaped by a past and committed toward a future, concerned for what becomes of us. We are not just subjects who are functionally equivalent to a program for processing representations. We do not occupy a position over against the world we survey and represent; we are situated within the world. Whereas theoretical hermeneutics took interpretation to be the formation of hypotheses, for

60. Heidegger 1957: 150, 1962: 191.
61. Heidegger 1977: 142, 17.

Heidegger everything we do is interpretive. This difference may seem to collapse if one recognizes that proponents of a theoretical hermeneutics may take our behavior to embody hypotheses we do not explicitly have in mind. Indeed, any account of these hypotheses will be underdetermined by our behavior, so that any description of them that accounts for what we do will be as good as any other. But for Heidegger, any such account of the world picture our behavior reflects will be equally *in*adequate; it will always miss what is *at issue* in what we do. This is because if there is to be anything at all at issue in what we do, it cannot be a belief or a value whose authority depends upon us.

The interpretation discloses what it is to be, rather than what is the case. It is not that this meaning is thematic in the interpretation; rather, this is the focus of our interpretation, what gives it sense and direction. Our understanding of what it is to be is exhibited in the style and the coherence (or incoherence) of what we do. We show it rather than say it. This is not to say that all examples show it equally well; some are more revealing than others because they provide a clearer focus to our sense of what it is to be us, albeit without doing so exhaustively or explicitly. This is the sort of claim Clifford Geertz makes about the cockfight as revealing what it is to be Balinese and that Heidegger makes about the Rhine hydroelectric station as disclosive of ourselves.[62] In any case, by Heidegger's account, what is disclosed in our interpretations and what is at stake in them are the same. But what is at stake is, to repeat, not a value(s) posited in the interpretation, but the place of the interpretation in its world.

The implications of Heidegger's practical hermeneutics for Dilthey's distinction between the study of nature and the study of human beings is not immediately clear. An influential tradition, of which Dreyfus and Taylor are leading exponents, suggests that if Heidegger is right, the study of human beings can never be "scientific" in the way physics or biology is, or at least can never become *normal* science.[63] This result supposedly occurs because any theory about human beings or their social interaction must leave something out—namely, what is at stake in what we do. I think this view is mistaken: what is left out is not a thing at all, and if the human studies are culpable for its omission, so are the natural sciences. We will not be able to see why this view is mistaken, however, until we have considered in more detail what we can make of the natural sciences if we adopt something like Heidegger's her-

62. Geertz 1979; Heidegger 1954, 1977.
63. For this last qualification, see Dreyfus (1984: 14).

meneutics. The sketch of the second interpretation of Thomas Kuhn gave us a first glimpse at what such a philosophy of science might look like. Now that we have seen the more general theory of hermeneutics it reflects, we are prepared to take a closer look at its implications.

Chapter 4

Local Knowledge

THEORETICAL hermeneutics in some ways represents a radical departure from received interpretations of science. Its anti-foundationalist verificationism and its ontological relativism coincide with and reinforce some of the most important developments in the critique of traditional empiricism, as we discussed it earlier. But there is an important respect in which this view expresses and vindicates a deeper tradition in the philosophical understanding of science. Theoretical hermeneutics is a theory-dominant philosophy of science. In the narrow sense, this characteristic means that it assigns a preeminent role to theories (i.e., a particular sort of semantic structure) within the practice of scientific research. Experiments and observations are significant only within a theoretical context. Theory guides the construction and performance of experiments, supplies the categories within which observations are to be interpreted, and mediates the transmission and application of the results of research. Ultimately, theories are the end product of research: the aim of science is to produce better theories.

There is also a deeper sense in which this understanding of science is theory-dominant, however. "Theory" has commonly signified a kind of understanding that is not tied to our practical involvements with the world. Charles Taylor has this in mind when he says: "A theoretical understanding aims at a disengaged perspective. We are not trying to understand things merely as they impinge on us, or are relevant to the purposes we are pursuing, but rather grasp them as they are, outside the immediate perspective of our goals and desires and activities."[1] But it is not just from our everyday concerns, interests, and involvements that such theoretical understanding is supposed to disengage us. Theoretical understanding is nonperspectival and therefore treats all locations in space or time as theoretically equivalent (it allows no epistemological privilege to any spatiotemporal framework). It likewise abstracts from all particular social contexts; this feature is the classical difference between theoretical knowledge, whose adequacy is invariant across varying speakers and forums, and rhetoric, which is aimed at a particular audience and evaluated in that context. Theoretical knowledge in turn bears no reference to a particular knower. Hesse's account of her learning-machine model of scientific development is thus squarely in the tradition in its indifference to whether it represents "the individual scientist, the whole body of scientists, or the institution of science represented by its learned societies, journals, and textbooks."[2] Note especially that it is not *an* individual scientist, but *the* individual (i.e., an abstract, representative shadow of a person) that the model might represent. The subject of such theoretical knowledge is supposed to be abstract and disembodied. Even the particular practices and techniques through which theoretical knowledge is discovered and validated are regarded as matters of accident with respect to the knowledge itself. What is important for the theoretical consideration of science is not the actual circumstances of the discovery and validation of its claims, but only a "rational reconstruction" of them. Finally, theoretical hermeneutics endorses the systematicity of scientific knowledge. It is no accident that Quine speaks of a "total theory" or the "total system of sciences."[3] From a theory-dominant point of view, knowledge must make up a consistent and coherent (and perhaps ultimately complete) whole. In our everyday practices, we readily employ in specific circumstances beliefs and practices that are inconsistent with other things we do. Such inconsistency does not by itself prevent the practical efficacy of our efforts, but from a theoretical perspective it must be evidence of some inadequacy in our knowledge. The notion that a scientific field must have a unified theoretical understanding, at least in the limited sense of its being internally consistent throughout, is essential to a theory-dominant view of science.

A theory-dominant view of science in this broader sense has important consequences for what is relevant to an "internal" or philosoph-

1. Taylor 1982: 89.
2. Hesse 1980: 125.
3. Quine 1953: 42–43, 1970: 5.

ical understanding of science. From a theory-dominant point of view the local site of investigation, the experimental construction, the technical facilities involved in that construction, the particular networks of social relations within which the investigators are situated, and the practical difficulties of getting on with research are incidental to scientific knowledge. Consider first the local site of investigation.

Scientific research is normally done at a specially prepared site, typically a laboratory or clinic, but sometimes a location in the field. From a theoretical point of view, however, this local environment is incidental to what is being explored or tested in the research. Scientific claims are universal. Any particular location merely provides an instantiation of these universal claims, and any particularity that must be taken account of is potentially an objection to the results we obtain. Thus, for example, we may study in a laboratory certain physiological effects on rats of intravenously injected somatostatin, but what we aim to study are these physiological effects *generally* (i.e., anywhere they might obtain), not as belonging to the laboratory. If the *particular* specifications of the laboratory as a local site of investigation must be taken account of in interpreting the results, it becomes possible to challenge those results as artifacts. Of course much special apparatus, perhaps unique to this particular facility, may be required to create this effect (somatostatin in buffered saline solution does not normally exist outside laboratories and laboratory supply houses, rats in the wild are rarely injected with anything, and the equipment used to record physiological variations may not exist anywhere else). But the experiment is carefully controlled to eliminate effects attributable to purely local contingencies, and the result is to be interpreted as an illustration of effects attributable to somatostatin however introduced, whether recorded or not, on any rat, whether laboratory bred or wild.

Experiments, then, are taken to be generally interpretable instantiations of theoretical claims. The results will (the experimenter hopes) count as a representative instance of universally quantified, nonindexical sentences describing and theoretically interpreting the physiological effects of somatostatin. Such sentences will of course normally be uttered on specific occasions, addressed to a particular audience (in journal articles written for professional colleagues, in review articles assessing results for a larger professional audience, in textbooks for students, in newspaper articles for lay readers, etc.), but their truth conditions are presumed to be invariant across these occasions of utterance. The particular antagonistic field within which the statement is uttered and accepted or rejected is subsumed to its evaluation within an abstract, unsituated field of statements. Hesse's learning machine is proposed as being itself a theoretical model for the evaluation of statements in this theoretical form (i.e., delocalized, shorn of particularity and indexicality, situated in a field of other equally eternal, delocalized sentences). When Quine talks about the "total system of science" or a "total theory," he is talking about a field of statements that can be treated in such an abstract, timeless, unsituated way. This field of statements is actually attached to experience only at the periphery, in a few carefully constructed, particular, local observations, but the particularity of these occasions is removed in taking them as particular instances of a universal theory. This theory could in principle have been instantiated to the same effect on a variety of other occasions, and it is therefore taken to refer to events (or possible events) that will not, and in some cases could never, be observed in the same way.

All of this account is commonplace and obvious. In this chapter I hope to make it seem less so. I will suggest an analysis of scientific practice that reveals the local, existential character of the understanding it produces. Scientific knowledge is first and foremost knowing one's way about in the laboratory (or clinic, field site, etc.). Such knowledge is of course transferable outside the laboratory into a variety of other situations. But this transfer is not to be understood in terms of the instantiation of universally valid knowledge claims in different particular settings by applying bridge principles and plugging in particular local values for theoretical variables. It must be understood in terms of the adaptation of one local knowledge to create another. We go from one local knowledge to another rather than from universal theories to their particular instantiations. The point is to give primacy not to particular occasion sentences, but to particular *occasions*—that is, to what we *do* (or can do) in particular situations. Even our knowledge of theories, I will argue, has to be accounted for in terms of such a practical, local grasp. "Theoretical understanding" in the sense that Taylor described it does not even account for our understanding of theories.

My account of science as local knowledge draws upon Kuhn's claim that scientific knowledge is embedded in the ability to employ concrete exemplars in the absence of agreed-upon interpretations of them; the new empiricist insight that the expansion of technical control in science is not dependent upon the particular developments of theoretical explanations of that control; and Heidegger's account of the importance of locally, materially, and socially situated skills and practices for all understanding and interpretation. I will also take account of recent microsociological investigations of the material and social setting of scientific research in laboratories. It is ironic that despite my reliance upon central features of Heidegger's hermeneutics, I must begin by challenging the specific account of scientific investigation Heidegger provides in *Being and Time,* for in this early work, Heidegger still relies upon a traditional theory dominance in assigning an ontologically distinctive role to science. My criticism of this role will launch our investigation of the local, existential character of science.

The Inadequacy of Heidegger's Early Philosophy of Science

To understand Heidegger's early philosophy of science, we must add one further consideration to our previous account of his practical hermeneutics. Heidegger's hermeneutics originally constituted a transcendental inquiry into the *meaning* of Being, which for him signified whatever was necessary for the disclosure of Being. More colloquially, he was asking how it is that anything shows up at all. His basic answer was that disclosure was possible only if the beings to which beings were disclosed had certain characteristics. Foremost among these was their belonging to a self-adjudicating community.[4] That is, they must recognize one another as belonging to that community, and this recognition must be based on the appropriateness of each one's behavioral responses to the shared environment. Those beings who do "what one does" are recognized as belonging; but "what one does" is discovered only through the behavior of those who belong. Alasdair MacIntyre effectively described this reciprocal recognition as the use of "schemata which are at one and the same time constitutive of and normative for intelligible action by myself and are also means for my interpretations of the actions of others."[5] Thus the authoritativeness of the behavior of those who count as belonging to the interpretive community is a necessary condition for disclosure. Heidegger adapts the term *Dasein* to refer to being such socially and behaviorally self-adjudicating interpreters.

Heidegger then claims that what there is apart from these socially self-adjudicating interpreters is revealed within the configuration of the practices they engage in.[6] The behavioral responses of those who count as belonging (as persons, we might say) determine what things

4. I am indebted to Robert Brandom (1983) for this felicitous phrase.
5. MacIntyre 1980: 54–55.
6. I am indebted to Mark Okrent for emphasizing that this represents a distinct stage in Heidegger's transcendental argument.

are. Now the most basic everyday way things show up through these responses is in their functionality. They are usable *for* something or other we engage in. This something or other in turn is what it is because of its place in relation to other things and practices. Mark Okrent has succinctly captured the contextuality of all such determinations.

> Thus, for example, whether an object is a hammer or not is determined by whether or not it is considered appropriate to pick it up and start hammering with it in a situation which calls for hammering. Now, of course, on the same account what is to count as hammering and what is to count as a situation which demands the response of hammering are equally going to be fixed by the acceptance of a given behavioral response, as indeed will "acceptance" and "rejection" themselves.[7]

The interrelations between hammers, the situations in which one hammers, the equipment one uses with hammers (nails, boards, etc.), and the purposes for which one hammers are internal relations: the things related do not exist as these sorts of things except in their actually belonging to such relations.

But Heidegger insists that such functionality, or "readiness-to-hand," does not exhaust the ways things can be other than as socially self-adjudicating interpreters (*Dasein*). Things can also be recognized and responded to as *de*contextualized. It is not that such things, which Heidegger calls "present-at-hand," exist independent of the behavioral responses of persons within a configuration of practices and functional equipment. It is that the appropriate behavioral responses to them are carefully shorn of any functional reference. Science, Heidegger thought, is the configuration of practices in which we decontextualize things from the configuration of everyday functioning. Electrons, DNA sequences, or pulsars are not recognized by their functionality. They are what shows up when we "just look at" these things in the appropriate way, where what counts as *appropriate* looking is still socially and behaviorally adjudicated. Heidegger thus characterizes science as a theoretical activity, disengaging us and its objects from functionality and locality in the ways described above. The important differences between this account and other theory-dominant accounts of science are that theoretical understanding is limited to science and, more important, that it is claimed to be derivative from and privative with respect to ordinary practical involvement. Thus Heidegger thinks he has described "the *ontological genesis* of the the-

7. Okrent, n.d.

oretical attitude"[8] out of our everyday practical dealings with the ready-to-hand.

Heidegger traces this ontological genesis of theory back to the momentary disengagement from the configuration of our practical involvements that occurs when equipment is broken, missing or in the way.[9] This "moment of truth" quickly dissipates, and we find ourselves reabsorbed in the practices of repairing, replacing, removing, or making do without the defective equipment. However, our dealings with it can also "change over" into a new way of looking at things as present-at-hand.[10] Heidegger is disturbingly vague about how this changeover is to occur; it is not at all clear in his account *how* one can get from a breakdown of practical involvement to the theoretical attitude. But once this happens, the ordinary functional contextuality of things gets *replaced* by the "mathematical projection of Nature."[11] By "mathematical," Heidegger means not "numerical" (broadly conceived), but rather the a priori projection of a *theoretical* conception that provides a new, afunctional context within which things can manifest themselves.

> [Modern physics] can proceed [to make use of a quite specific mathematics] only because, in a deeper sense, it is already itself mathematical. *Ta mathemata* means for the Greeks that which man knows in advance in his observation of whatever is and his dealings with things. . . . Only because numbers represent, as it were, the most striking of always-already-knowns, and thus offer the most familiar instance of the mathematical, is "mathematical" promptly reserved as a name for the numerical.[12]

Things taken to be present-at-hand, having lost the intelligibility arising from their belonging to a functional configuration of everyday concerns, would be unintelligible in isolation,[13] but they acquire a new, delocalized lucidity within the context of scientific theories. Thus Dreyfus seems to follow the early Heidegger when he suggests that something like a post-Quinean theoretical hermeneutics appropriately characterizes the development of scientific understanding but cannot be considered as a general account encompassing our everyday grasp of things.[14]

8. Heidegger 1957: 357, 1962: 408.
9. Heidegger 1957: 72–76, 357–64, 1962: 102–7, 408–15.
10. Heidegger 1957: 361, 1962: 412–13.
11. Heidegger 1957: 362, 1962: 413–14.
12. Heidegger 1952: 72, 1977: 118–19.
13. Heidegger 1957: 360–64, 1962: 412–15.
14. Dreyfus 1980.

To understand what is wrong with this account, we must consider the place of practical involvement in the supposed decontextualization of things present-at-hand and their recontextualization in the mathematical projection of Nature. Heidegger himself notes that "theoretical research is not without a praxis of its own,"[15] but his understanding of the scope of such praxis considerably restricts it. As examples of research praxis, he cites the technical design and setup of experiments, the preparation of microscope slides, and archaeological excavation. He clearly thinks these are examples of research practices that are only *associated* with theoretical cognition, because he goes on to say: "But even in the 'most abstract' way of working out problems and establishing what has been obtained, one manipulates equipment for writing, for example."[16] What is conspicuously absent in these choices of examples is any suggestion that scientific knowledge itself involves any practical circumspection. Scientific knowledge on this view is a form of disengaged viewing guided by a theoretical ("mathematical") projection. Experiment (and its associated practice) can illustrate and validate our cognition of the present-at-hand, but the latter is still a distinct way of comporting ourselves toward the world, derived from practical engagement but shorn of its specifically practical character.

There are three features of scientific research that Heidegger cites to reinforce his claim that it involves a decontextualized "viewing" of objects present-at-hand. The first is that science thematizes its objects. Heidegger in his account of the readiness-to-hand of equipment emphasized that when it functions properly, equipment recedes from explicit consideration.

> The peculiarity of what is proximally ready-to-hand is that, in its readiness-to-hand, it must as it were, withdraw in order to be ready-to-hand quite authentically. That with which our everyday dealings proximally dwell is not the tools themselves. On the contrary, that with which we concern ourselves primarily is the work—that which is to be produced at the time.[17]

The objects we deal with in science, by contrast, are our thematic concern. Our aim is to discover what and how they are, not to take advantage of what they are for. It is not their serviceability within a configuration of practices, but their properties as disclosed within a

15. Heidegger 1957: 358, 1962: 409.
16. Heidegger 1957: 358, 1962: 409.
17. Heidegger 1957: 69, 1962: 99; see also Ihde 1979, chaps. 1–4.

mathematical projection that concern us. Practical involvement directs us away from things toward a larger configuration of concerns, whereas theory directs us back toward the thing as it stands before us.

The second feature Heidegger finds decisive for the difference between scientific theorizing and practical dealings with things is the delocalization of the objects of science. Heidegger points out:

> In the 'physical' assertion that 'the hammer is heavy' we *overlook* not only the tool-character of the entity we encounter, but also something that belongs to any ready-to-hand equipment: its place. Its place becomes a matter of indifference. This does not mean that what is present-at-hand loses its 'location' altogether. But its place becomes a spatio-temporal position, a 'world-point', which is in no way distinguished from any other.[18]

In our everyday practical dealings, things belong in places that thereby become significant as places. The hammer is "on the shelf," "with the nails," or "by my side"; it can also be "out of place" or "misplaced." From a scientific, theoretical point of view, the significance of places is dissolved, and with it the belonging of things to places. This detachment is what Heidegger means by places becoming positions. Positions in a mathematical frame of reference have no differential significance. They also have no center, no orientation, no directionality, and no internal interconnections. All the delineations with which spatiality is structured by our practical dealings with things are removed from scientific, theoretical consideration. Above all, the locale where we live, and within which our concerns are focused, becomes just another represented position, not distinguished from any other. Theoretical representation is indifferent to local situations.

Finally, Heidegger emphasizes the transformation of language that occurs in science. Science aims to produce assertions stripped of all indexicality. Indexicality is usually introduced relative to an observer or speaker. Heidegger, however, is concerned with the implicit indexing of a task or situation. In his account of the genesis of the theoretical attitude, he points out the transition that can be made from contextually situated claims like "the hammer is too heavy" (for the task at hand) to "the hammer is heavy" (a decontextualized assertion of a contextual property) to "the [specified] hammer has a mass of 1.2 kg." And of course we usually go further than this in science, removing even specified particularity by making universally quantified asser-

18. Heidegger 1957: 361–62, 1962: 413.

tions. Heidegger notes that "when this kind of talk is so understood, it is no longer spoken within the horizon of awaiting and retaining an equipmental totality and its involvement-relationships. . . . [The] entity in itself, as we now encounter it, gives us nothing with relation to which it could be 'found' too heavy or too light."[19] The disclosure of a thing as a massive body in a gravitational field seems to be disconnected from its functional references to a local configuration of practical involvements.

Heidegger's claim seems to be reinforced by the preeminence of written documentation in science. As Bruno Latour and Steve Woolgar noted in their anthropological study of a scientific laboratory,

> it seemed that there might be an essential similarity between the inscription capabilities of apparatus, the manic passion for marking, coding, and filing, and the literary skills of writing, persuasion, and discussion. Thus, the observer could even make sense of such obscure activities as a technician grinding the brains of rats, by realizing that the eventual end product of such activity might be a highly valued diagram. . . . For the observer, then, the laboratory began to take on the appearance of a system of literary inscription.[20]

This analogy is important because, unlike speech, which vanishes with the occasion of its utterance, inscriptions persist and show up and are read in situations far removed from their writing. Nor is this feature accidental. Samuel Todes has pointed out that inscriptions are the "visual substantial trace" of

> sensuous characters, [which] compared to natural shadows are *radically distinct* in their mode of representation. They have an absolute, a universally univocal, sense in place of the contextually variant and analogical sense of natural shadows. This is due to the character of the blank page (or other blank inscribed surface) which serves as a uniform and universal, rather than variegated and local, setting in which sensuous characters can appear.[21]

Scientific research, with its various and complicated activities in the laboratories, aims toward an outcome that can be shorn of all reference to the particularities of its production and situated in a context of representations (i.e., other inscriptions) that can be made entirely

19. Heidegger 1957: 361, 1962: 412.
20. Latour and Woolgar 1979: 51–52.
21. Todes 1975: 111.

thematic because its background (the page) has been emptied of all content.

Heidegger's description of science, as the decontextualizing discovery of things present-at-hand and their subsequent thematic recontextualization within a theoretical projection, seems initially compelling. Why, then, do I reject it as fundamentally inadequate? I will argue that in this early work Heidegger conceded too much to the traditional theory-dominant understanding of science. Heidegger can describe science as decontextualizing, and can take the theoretical ("mathematical") projection of nature to be a form of disclosure fundamentally different from everyday practical dealings with things, because he fails to give close consideration to the actual practices involved in scientific research. What Heidegger takes to be a phenomenological description of the theoretical attitude, I will argue is laden with theory-dominant prejudices imported uncritically from the tradition. It is perhaps no accident that Heidegger's early analysis of science is the one central position he takes that accepts wholesale the theory-dominant investigations of his mentor Edmund Husserl.

Specifically, I will challenge Heidegger's account, and the theory-dominant tradition it reflects, on four principal points:

1. Heidegger misunderstands the practices of theoretical representation themselves: he overlooks the horizons within theories themselves (i.e., as distinct from situating the entire theoretical project within the horizon of our practical dealings with things), and he neglects the ways theories are used, their concrete binding to practical research situations, and the practical, circumspective, opportunistic character of the reasoning they involve.

2. The role of experiment in science, and with it the significance of both the local, material setting of the laboratory and the technical and practical know-how developed within it, are correspondingly misunderstood; in this case the theory-dominant perspective that Heidegger still retains here reduces experiment to a merely incidental practice in science; it is not even clear why science typically is practiced in laboratories and clinics.

3. What Heidegger characterizes as theoretical decontextualization is better construed as a process of standardization, whereby scientific objects and practices become reliably transferable to new research contexts. What Heidegger takes as a disengagement from particular practical situations and interests is actually a consolidation of power, which enables power to be extended into new networks.

4. Heidegger does not take seriously enough his own insistence

that any being and its determinations can be manifest only to a behaviorally self-adjudicating community. As a result, he fails to see the social, functional contextuality that still governs the acceptability of the most abstract theoretical claim.

Theory as Practice and Research as Action

Let us begin by reconsidering the character of scientific research as an activity. Early in his discussion in *Being and Time* of the ontological genesis of science, Heidegger notes that his concern is to grasp the "existentially necessary [conditions] for the possibility of *Dasein*'s existing in the way of scientific research,"[22] and he contrasts such an "existential conception of science" to a retrospective logical analysis of the results of scientific research. It is thus ironic that Heidegger actually has very little to say about scientific research as something we do (an activity). His discussion of science as a decontextualized, mathematical projection of the natural world focuses on the existential conditions of theoretical cognition. There is almost no attempt to consider how such cognition might be developed and extended or how difficulties within it might be recognized and resolved. Heidegger does eschew any central concern with the retrospective analysis and justification of the finished products of research activities. But his discussion of science *is* focused upon the interpretation of finished results and upon the relation between such interpretation and what takes place in everyday circumspective concern. The issue he takes up is not how we can do research in molecular biology or high-energy physics but how we come to comprehend general theories (of which Heidegger takes classical mechanics as illustrative).

The most widely known attempt to distinguish between the image of science one gets from regarding research and the one that results from looking at finished theories was developed by Thomas Kuhn, drawing extensively upon an earlier discussion by Ludwik Fleck.[23] Kuhn began by pointing out that

> the image of science by which we are now possessed has previously been drawn, even by scientists themselves, mainly from the study of finished scientific achievements as these are recorded in the classics and, more recently, in the textbooks from which each new scientific generation learns to practice its trade. Inevitably, however, the aim of such books is

22. Heidegger 1957: 357, 1962: 408.
23. Kuhn 1970a; Fleck 1979. Kuhn acknowledges his debt to Fleck in the preface to his book.

> persuasive and pedagogic; a concept of science drawn from them is no more likely to fit the enterprise that produced them than an image of a national culture drawn from a tourist brochure or a language text.[24]

Kuhn insisted that "textbook views" misleadingly reconstruct the development of science as a steady accumulation of results and the evidence to support them. He thinks they overlook the conceptual reformulations and the shifts in interests and standards that have marked the actual history of science. Fleck perhaps goes further by also insisting that one thereby overlooks the social process by which the conflicting, personalized, and idiosyncratic methods and viewpoints of particular research workers are reformulated into a single, collective, consistent account.[25]

Heidegger's (and implicitly, Quine's or Hesse's) discussion misunderstands the activity of research in more subtle ways. They would generally agree that the history of science cannot be adequately construed as a cumulative process. Nevertheless, it is assumed that at any given time research work presupposes a background of unified theoretical understanding. This theoretical background is what Quine calls our "total theory" or "the total system of science," what would be represented in Hesse's model by the state of the learning machine at any given time and what Heidegger speaks of as "the mathematical projection of Nature." This background is now recognized as constantly shifting. It touches upon experience only at selected points, may be riven by unseen inconsistencies, and in any case is probably never fully articulated as a whole. Nevertheless it functions as a single interconnected theory or "mathematical projection," providing the background against which research projects become intelligible and within which our justifications acquire force.

Quine does not usually argue for the systematic character of background theory, although it is clear that considerations of consistency are of foremost importance in his insistence upon it. This emphasis occurs because of the importance of truth for Quine as the determinant of any interpretation. Theories aim at truth, the determination of what is (accepted as) true is prior to any possible determination of meaning, and consistency is essential to truth. If a particular theory cannot be consistently linked to one's total theory, both cannot be true. Hence the importance of the systematic character of science.

24. Kuhn 1970a: 1.

25. If Kuhn differs from Fleck here, it is in saying that this processing primarily takes place in the training of scientists, before their full-scale participation in scientific research.

For Heidegger, our understanding need not be entirely linked in this way as a systematic whole; indeed, he thinks, it cannot be so. Nevertheless, he thinks that the mathematical projection of nature must form a single, consistent whole. This has to do with the representational character of such projection. The local "world" within which we find ourselves amid meaningful things (equipment) and intelligible practices has been decontextualized and replaced by a projected theoretical world picture. Such a theoretical context does not have the intelligibility born of socialization, familiarity, and functionality. The bestowal of meaning usually achieved via the "practical holism" of our belonging to a world must instead be accomplished by the systematicity we put into our theories.

> Here [in modern representing, in distinction from Greek apprehending] to represent means to bring what is present at hand before oneself as standing over against, to relate it to oneself, to the one representing it, and to force it back into this relationship to oneself as the decisive realm. Wherever this happens, man "gets into the picture" in precedence over whatever is.[26]

> "We get the picture" concerning something does not mean only that what is, is set before us, is represented to us, in general, but that what is stands before us . . . as a system.[27]

We thus understand *only* the systematic, "totalizing" relations we have put into things as represented, because we have methodically stripped away any other intelligibility they might have.

When the background against which research takes place is thus conceived to be a systematic theory, the metaphor of "problem solving" seems particularly appropriate to characterize the activity of research. Consider Quine's metaphor of a theoretical field bounded by experience at the edges. The task of research is to refine the theoretical structure (and the experience it permits us to describe) to improve the fit at the boundaries. Research tasks can then be neatly categorized as either empirical or conceptual problems, as Larry Laudan has done.[28] Empirical problems are "anything about the natural world which strikes us as odd, or otherwise in need of explanation."[29] As Laudan immediately points out, however, "our theoretical presuppositions about the natural order tell us what to

26. Heidegger 1952: 84, 1977: 131.
27. Heidegger 1952: 82, 1977: 129.
28. Laudan 1977: 15, 48–49.
29. Laudan 1977: 15.

expect and what seems peculiar or 'problematic' or questionable (in the literal sense of that term). . . . Hence, whether something is regarded as an empirical problem will depend, in part, on the theories we possess."[30] Empirical problem solving is thus primarily the attempt to resolve conflicts or omissions at the boundaries of our theories, to expand the *systematic* scope of our theories. Conceptual problems are even more dependent upon the requirement of systematicity. Laudan describes two basic kinds of conceptual problems: internal inconsistencies exhibited by a single theory, and inconsistencies or incompatibilities between two theories external to one another. Laudan, however, is not wedded to the notion of science as implicitly embodying one total or systematic theory, as are Quine and early Heidegger. If we combine his account of problems with their general understanding of science, what Laudan calls "external conceptual problems" are actually internal to the total system of science. The image of scientific research that results is that it is the resolution of conflicts within or at the boundary of systematic theory.

A more adequate alternative to the understanding of scientific knowledge as a theoretical system and research as the resolution of conflicts and embarrassing omissions within that system begins with a rather different understanding of scientific theory. Instead of being seen as a network of interconnected sentences or conceptual schemes, theories are taken to be extendable models. The locus classicus for this interpretation is once again Thomas Kuhn's *Structure of Scientific Revolutions,* with interpretive emphasis placed upon his account of paradigms as exemplary problem solutions. Kuhn argued that theories were not primarily sentential systems whose representational content was first learned and only then applied to specific situations, perhaps with the help of additional bridge principles. He claimed instead that the content of theories was embedded in standard, exemplary solutions to model problems. Learning the theory is learning to understand these problem solutions in such a way that what was done in the model case can be extended and transformed to deal with a range of more or less similar cases. Theories on this view are tools one learns how to use rather than sentences whose implications one comes to know. Instead of being systems of sentences whose applications are deductively derivable, theories are a loosely connected set of models extendable by analogy. One increases one's grasp of the theory by moving from one concrete case to another rather than from theoretical generalization to specific application.

30. Laudan 1977: 15.

> The student discovers, with or without the assistance of his instructor, a way to see his problem as *like* a problem he has already encountered. Having seen the resemblance, grasped the analogy between two or more distinct problems, he can interrelate symbols and attach them to nature in the ways that have proved effective before. . . . The role of acquired similarity relations also shows clearly in the history of science. Scientists solve puzzles by modeling them on previous puzzle-solutions, often with only minimal recourse to symbolic generalizations.[31]

Kuhn's claim has recently been picked up and extended by Nancy Cartwright. She follows Kuhn in insisting that the content of a theory is contained in model treatments of specific problems.

> On the simulacrum account, models are essential to theories. Without them, there is just abstract mathematical structure, formulae with holes in them, bearing no relation to reality. Schroedinger's equation, even coupled with principles which tell what Hamiltonians to use for square-well potentials, two-body Coulomb interactions, and the like, does not constitute a theory *of* anything. To have a theory of the ruby laser, or of bonding in a benzene molecule, one must have models for those phenomena which tie them to descriptions in the mathematical theory. In short, on the simulacrum account, the model is the theory of the phenomenon.[32]

She goes beyond Kuhn, however, in two important respects. She has argued that the number of model problem solutions must be constrained if they are to constitute a theory. Otherwise, we would have not a theory but an ad hoc collection of problem treatments. Thus, even though theories grow by gradually developing new treatments of related problems, this growth must considerably outpace the growth in the number of models required to achieve it.

> At heart the theory works by piecing together in original ways a small number of familiar principles, adding corrections where necessary. This is how it should work. The aim is to cover a wide variety of different phenomena with a small number of principles, and that includes the bridge principles as well as the internal principles. It is no theory which needs a new Hamiltonian for each new physical circumstance. The explanatory power of quantum theory comes from its ability to deploy a small number of well-understood Hamiltonians to cover a wide range of cases.[33]

But Cartwright thinks this quest for explanatory power requires that the theoretical models we employ be rather remote from any actual

31. Kuhn 1970a: 189–90.
32. Cartwright 1983: 159.
33. Cartwright 1983: 139.

empirical situation. As Cartwright notes, in texts on quantum mechanics "generally there is no word of any material substance. Instead one learns the bridge principles of quantum mechanics by learning a sequence of model Hamiltonians. I call them 'model Hamiltonians' because they fit only highly fictionalized objects."[34] Theory in physics, she thus claims, is embodied in a limited number of model mathematical treatments of idealized situations. These models rarely are descriptively accurate for any actual empirical phenomenon. Nevertheless, with various practical adjustments and distortions, we can deal with many real situations, often quite complicated, by treating them as variations on the model.

One last point must be added to this initial sketch of an alternative conception of the scope of theory. Kuhn and Cartwright challenge the traditional view of theory as a unified, systematic deductive structure. They present us with a picture of theory as a disjoint collection of models whose range of application is not fully specified and whose effectiveness and accuracy vary considerably within that range. Furthermore, the use of theory is more a practical matter of making ceteris paribus adjustments to a concrete model to fit the needs of a specific case on a particular occasion than it is providing a formal, deductive derivation from general principles. Ian Hacking takes us one step further along this line in his discussion of the kinds of theoretical principles employed in science. Theory, he tells us, is not just one thing with one function. Theories may be sweeping, suggestive speculations that only point us in a general direction, or they may be elegant formal mathematical representations. They may be physical models that give us a qualitative understanding of causal interactions or more or less ad hoc mathematical representations (what Cartwright calls phenomenological laws).[35] One does different things with theories of these various types. They do not form a seamless web of belief. Their coverage overlaps, and they may provide inconsistent versions of the same phenomenon. Some phenomena may fall in the gaps between the various kinds of theory we have in a domain and are consequently not well treated by any. Theory gives us not one "world picture" but a diverse range of representations and manipulations. Scientific theory provides not one kind of thing we believe, but many kinds of things we do.

We can now return to the previously mentioned account of research as problem solving. The image of a "problem" implies a con-

34. Cartwright 1983: 136.
35. Hacking 1983: 212; Cartwright 1983, passim.

flict of some sort: an inconsistency within theory that should not exist, a conflict between theoretical expectation and empirical result, an inability to predict or explain some phenomenon that nevertheless seems well within the presumed capabilities of our theories. This image fits well with the conception of theory as aiming to provide a systematic representation of the world; inconsistencies are unacceptable difficulties, predictive failures call for revision in theory, and significant omissions represent insufficiencies in the scope of our theories. But it does not fit so well with the image of theory suggested by Kuhn, Cartwright, and Hacking. Not all conflicts between theory and empirical results are objectionable, since some clearly reflect peculiarities of the particular empirical situation (and its description or construction, or both) rather than substantial flaws in the theory. Only if the difficulty suggests analogous problems in handling other situations, or if this particular situation has some intrinsic interest or some instrumental role in further research, will such conflicts literally be problems. In such cases we have to ask why the difficulty *needs* to be resolved before its presence can count as an objection to using the theory in other ongoing research.

Of course, many current issues in science straightforwardly fall into the "problem" category. But even then it is useful to examine how such problems develop. Consider the following current problem in nuclear astrophysics. Experimental procedures designed to detect the solar neutrino flux now register a value approximately one-third that predicted by the best theoretical models of thermonuclear reactions in the sun's interior.[36] This result is clearly regarded throughout the field as a problem that needs to be solved, since models of stellar interiors are sufficiently speculative that any detailed empirical articulation of them is a significant focus for further theoretical development. A variety of proposals have been made, ranging from criticisms of the experimental design to significantly revised theories of stellar interiors, and most astronomers believe that a variation on one of these approaches will eventually be able to resolve the discrepancy. This issue looks like a classic case of an empirical "test" of theory and a concerted effort to solve the problem generated by this test. But it is important to recognize that the original work that generated the "problem" had a rather different motivation than this account of it suggests.

At the time of the first proposed experiments to detect solar neutrinos (the mid-1950s), it was generally believed that the solar neu-

36. William Herbst told me this.

trino flux was too small to be detected with current technology. The initial impetus for such experiments came from an experimenter who had developed the equipment and techniques and had trained technicians to detect neutrino interactions. Having devoted considerable resources to developing a neutrino detector, he was looking for neutrinos to detect, and he was trying to enlist some theoreticians to provide an expected value for the solar flux that he could use as a target.[37] Changes in the accepted view about stellar nuclear reactions suggested to several theoretical astrophysicists in 1958 that the solar neutrino flux might be just sufficient to be detectable. By 1962 this possibility, based upon the availability of a sensitive neutrino detector, led several such scientists to collaborate on the complicated calculation of just what the expected value for the solar neutrino flux might be. What is important to recognize is that there was no intrinsic theoretical interest in solar neutrinos; indeed, those scientists whose expertise was required to produce models of the sun's inner structure and nuclear reactions "could see little point in working out the Sun in the detail required and were more interested in the late stages of stellar evolution."[38] What made the calculations significant was simply the availability of the apparatus needed to test their result.[39] Of course once the calculation was made, the experiment performed, and the basic reliability of both accepted by the astrophysics community, there was then a problem. But the original reason for the project was not a problem but an opportunity.

This distinction is not insignificant. Much scientific research is occasioned not by the felt need to resolve known difficulties in current theory, nor by the desire to uncover such difficulties,[40] but by the concern to take advantage of the available resources in equipment, techniques, trained personnel, and related scientific results.[41] The relevant question is what we can do with what we have, not what we must do in order to reduce conflict within current theory. Problems of the kind Laudan described do exist, of course, and they provide a significant part of the research opportunities available at any time. A known difficulty that seems to be within the scope of one's abilities to resolve constitutes a significant opportunity. But there are clearly

37. Pinch 1980: 80–84.

38. Pinch 1980: 88.

39. John Bahcall, quoted in Pinch (1980: 87). See also Kuhn 1970a: 26–29.

40. Kuhn 1970a: 52. Kuhn's claim that normal science does not *aim* to produce significant novelty is reflected in Pinch's interviews with physicists involved in the neutrino detection experiments; Pinch 1980, passim.

41. Knorr-Cetina 1981: 33–47; Latour and Woolgar 1979: 151–83.

significant research opportunities that one must strain to regard as problems in this sense.

The importance of this distinction becomes clearer if we consider the kinds of project assessments required by an opportunistic conception of research compared with a problem-solving model. Determining what constitutes a problem within currently accepted theory involves probing for hidden inconsistencies, constructing novel experimental tests of the theory's implications, and assessing which areas of its domain are not adequately known. These are tasks that can plausibly be regarded as theoretical in Taylor's sense. Determining the outstanding problems is separable from the practical decision of which ones are worth working on with one's available resources. The former assessment is a relatively context-free analysis of the state of current theory, whereas the latter is bound to particular local situations. But an opportunistic conception of research undermines this distinction. What constitutes a research opportunity is not specifiable apart from a consideration of locally available resources and needs. There can be no opportunities in abstraction from the concrete situations where they arise, in the way that problems can be specified independent of the particular situations within which they might plausibly be solved. To the extent that apparent problems do exist, we might think they could still be identified "theoretically"; but even this limited claim is not acceptable, because not all theoretically identifiable problems constitute research opportunities. If no one is prepared to work on them, whether for lack of resources, interest, and collaborators or because they do not now seem amenable to solution, then these problems will not figure in current research.

It should be clear that the assessment of research opportunities constitutes what Heidegger would call circumspective concern. It is a practical assessment of what it makes sense to do, given the resources available and the aims and standards that govern scientific practice within a given field. Much of what one considers in such circumspection is the current state of knowledge: what results are (and are not) reliable bases for further work, what tools and techniques are sufficiently precise and illuminating, what prospective achievements would constitute a significant advance and what would not. Even here, however, one's assessment is an involved, practical one. In assessing previous results, one is assessing the credibility of the scientists who achieved them as much as the degree of confirmation of the results themselves.[42] One also relies on one's own "craft knowledge" of the

42. Latour and Woolgar 1979: 154–66, 194–208.

techniques and procedures involved and on one's own way of approaching research tasks.[43] Even in deciding what would count as an advance, one utilizes to a significant extent locally accepted assessments of where the field as a whole is going. After all, one will see one's own work as more central to the field than it will seem to many others, and this reflects a different grasp of what is going on in the field.

One must consider which techniques and procedures are appropriate to a project not only generally, but also in one's local context. Not all usable and otherwise appropriate techniques and procedures can be adequately performed with the facilities and expertise one already has available or can readily acquire. Nor is this limitation the only local factor that enters into such assessments. Projects differ in how long they might plausibly require to show results; particular scientists and laboratories differ in how long they can afford to wait. Some projects involve greater risk that no significant results will be obtained; whether such problems constitute research opportunities requires considering what risks one can afford to bear. All these decisions are affected by one's competitive situation. Are others likely to work on the same problem? What resources do they have, and how competent are they? By taking a different approach than your competitors, can you better protect yourself against the possibility of being preempted? And so forth.

The ways these local considerations may come into play can be illustrated by a perhaps too familiar example. Consider the research groups working on the structure of deoxyribonucleic acid in the early 1950s. From the perspective of the structural chemists who might typically be expected to take up this project, DNA was just another molecule. W. T. Avery's 1944 publications indicating that DNA was the principal material basis of heredity had had significant impact among biologists but were perhaps not so well appreciated elsewhere. To grasp the potential of DNA as a research opportunity and act on it required the technical skills of X-ray crystallographers and structural organic chemists and a grasp of the significance of the possibilities for molecular studies of genetics that was most clearly displayed by the new group of geneticists (often physicists turned geneticists) working with bacteria and bacteriophages. To many in this "phage group," after all, DNA was now *the* most biologically interesting molecule. But there were very few places in the world where this configuration of skills and interests existed.

The principal research group whose ongoing investigations focused

43. Ravetz 1971, chap. 3; Knorr-Cetina 1981, chaps. 1–2.

on structural studies of DNA was the King's College (London) group of Maurice Wilkins, Rosalind Franklin, and Raymond Gosling. Their relevant capabilities were in X-ray crystallography. They had acquired substantial technical skill and experience at taking X-ray photographs of DNA and were beginning to show promising results. Aside from this being their own strength, they saw X-ray studies as the only really legitimate and useful way to investigate the structure of large molecules. X-ray crystallography was then the only technique that could produce direct evidence for such structural determinations. Considerations from the theory of chemical bonding (themselves influenced by previous X-ray studies) provided constraints on proposed structures, but only X-ray data indicated positively what the structure was. There was the further influence of the prestige accorded to "hard" techniques like X-ray crystallography: they incorporated a great deal of physical theory that was considerably better established than the claims under investigation, and once results were achieved they offered less leeway for alternative interpretations.[44]

James Watson and Francis Crick at the Cavendish Laboratory in Cambridge understood the situation rather differently. They themselves had neither the facilities, the expertise, nor the time to do X-ray studies, although Crick was quite capable of interpreting X-ray evidence. They (especially Crick) were not even permitted to devote full time to the investigation. But they believed that their biological training (especially perhaps Watson's, whose graduate work was under phage group pioneer Salvador Luria) made them better situated than Wilkins and Franklin to appreciate the biological importance of the project. They also felt that enough was now known about the characteristic features of DNA to severely constrain its structural possibilities. Finally, at the Cavendish they had ready informal access to persons who had the expertise they lacked in experimental X-ray work and structural chemistry. These circumstances all made it seem plausible to attempt to incorporate the known constraints into physical models that could then be put together rather like jigsaw puzzles. This approach was a high-risk technique. Whereas Franklin would produce useful, publishable X-ray data whether or not she was successful in determining the structure, Watson and Crick's work was virtually useless unless it actually yielded the structure or conclusively ruled out other plausible alternatives. But they were in a much better position to take risks. It was not their principal research activity, at

44. Latour and Woolgar 1979: 142, 146.

least officially, and they were in a situation in their careers where they did not need to show immediate results from this work.

Both groups' deliberations were informed by their understanding of their competitive milieu. Franklin and Wilkins saw themselves in an essentially noncompetitive situation. They did not seem to have the biologists' sense of the urgency of the project; they did not think the Cambridge group was even working on it; and they did not take seriously the prospects either of Cambridge or of Linus Pauling's group in California, because they thought neither group had access to the vital X-ray data they had assembled and because they put little stock in the practical prospects of the modeling techniques either group would have to use. In the absence of competition, it was understandably much more prudent to proceed methodically and get it unquestionably right than to undertake risky methods that offered (to them) little practical hope. Crick and Watson, by contrast, were impressed with the immediate competitive threat they perceived in Pauling's work and clearly believed that only the "quick and dirty" procedure of physical modeling offered any real competitive opportunity.

Once the project was undertaken, these two approaches required rather different practical assessments of the reliability of available information. This issue was a much more serious concern in Cambridge, where it was crucial to include every available known constraint upon the structure but to avoid being misled by artifacts. Incorporating unreliable constraints could lead them far astray, but too much skepticism would leave their modeling seriously underdetermined. Also, virtually every important bit of information they had to rely upon had been developed by others and in some cases was not reported with sufficient accuracy or completeness. Determining what was a reliable result was a central issue.[45] The possibilities for going wrong were particularly illustrated by an earlier paper of Pauling's, which took account of much known structural information but overlooked both the water content of the samples whose photographs he worked with and a basic feature of the known chemistry of the molecule.[46] The result was an embarrassing failure. Franklin, by con-

45. Such problems explicitly arose over how seriously to take their report of Franklin's data, for example, concerning the water content of her samples and the two structural forms of the molecule; Chargaff's rules concerning the apparently equimolar ratios of adenine/cytosine and guanine/thymine, which many other scientists thought to be an artifact; the "standard" textbook structures for guanine and cytosine; the amount of variation possible in standardly determined bond angles; and so forth.

46. Cohen and Portugal 1977: 253–54.

trast, was primarily concerned with the reliability of her own previous investigations and technical decisions. Later in this chapter I shall discuss the practical difficulties of determining when one has achieved reliable or certifiable empirical results. The point for now is that Franklin's assessment was of a rather different kind than the secondary criticism of the literature that Watson and Crick engaged in. She did not have to consider whether she had confidence in the capabilities of others or whether their work had been accurately and completely reported. Furthermore, she could afford to be agnostic about such claims as Chargaff's rules. Her work did not rely upon them, and indeed it would eventually be part of the basis for judging the accuracy of Chargaff's work. Watson and Crick, by contrast, needed to decide at the very outset of each modeling attempt whether to use Chargaff's claim, for their next steps followed from this decision among others.

I have been arguing that there can be no "theoretical" or "objective" assessment of what counts as a research opportunity. The point is not that such determinations are irrational or subjective, but that the reasons for such assessments cannot be usefully disentangled from one's involved, skillful "craft knowledge" of a field of objects and practices or from one's practical needs—what one aims to do and hopes to achieve. In both respects this understanding is locally situated; reference to a particular configuration of persons, skills, equipment, and concerns is always crucial to it. To put this point in Heidegger's terms, the interpretation of opportunities in research requires circumspective possession of *Vorhabe* and *Vorsicht.* But it might be argued that one could also consider what is an opportunity for anyone, or at least one whose specification presupposes no one particular situation. Such considerations would, however, abstract only partially from a local understanding. One might, for example, have noted (as I have done) that modeling the structure of DNA would have been an opportunity for anyone who had knowledgeable access to enough of the available X-ray and chemical data (and the relevant theory) and a good machine shop, and who had some understanding of the project's biological significance. But even this supposedly unsituated account severely constrains the possible situations within which such an opportunity would have been present, and its very specification presupposes either a particular local grasp of comparable possibilities or a retrospective assessment based upon the actual achievement. This local understanding would provide the necessary grasp of the knowledge, practices, and equipment referred to and would above all give the needed basis to be able to specify what might count as *enough* data, *adequate* understanding, a "*good*" machine shop, and so forth.

A sympathetic critic might at this point in my argument concede the preceding claims in order to fall back upon some version of a distinction between the context of discovery and the context of justification. The assessment of what is an opportunity may be practical, engaged, and local, for this affects only how one goes about discovering new knowledge (or mistaken claims to it). But the evaluation of the results of our pursuit of such opportunities must be disengaged and objective. De facto, people in different situations may disagree about questions of justification, but there must be an internal objective to which they may appeal. Thus, for example, Franklin and her co-workers had no difficulty recognizing the success of Watson and Crick's work, despite their doubts about the procedures that led to it.

Unfortunately, such a distinction will not do. The principal reason is that the criteria for evaluating the adequacy of a research result depend directly upon the uses to which that result might be put. It is not just that a result that will serve as an accepted basis for further research must be more carefully checked to avoid future blind alleys. It is that the requirements of future use will determine how much precision is required and will often dictate the form of acceptable results as well. What these requirements are will in turn depend upon what is taken as an opportunity within various local configurations of facilities, expertise, funds, interests, and so on. Thus the local, practical, engaged understanding within which opportunities become manifest also determines in part what can count as a justified scientific claim. Later in the chapter we shall consider a more extensive argument for the circumspective character of justification. For now an example will clearly illustrate the way practical considerations of how one might proceed with research can drastically affect what counts as an acceptable result.

In the early 1950s Geoffrey Harris had proposed that the secretion of hormones from the pituitary gland was regulated not by nerve impulses, but by other hormones secreted by the hypothalamus. To investigate this hypothesis, initially one needed a reliable assay method to detect the presence of a pituitary hormone. One then needed to culture pituitary cells to determine that they would not by themselves secrete hormones in vitro. Adding hypothalamic tissue to the culture should cause the pituitary cells to resume hormone synthesis and secretion if Harris's proposal was correct. This effect was successfully demonstrated in 1955. It is a long step, however, from knowing that something in or from the hypothalamus regulates pituitary hormones to ascertaining what that something is and how it works. At this point many research strategies are possible, and they exert different requirements upon one's methods and results. The initial strategy chosen by most groups in the field was to investigate further the physiological effects of these postulated hypothalamic hormones (called "releasing factors"). This strategy required acquiring, processing, and fractioning hypothalami, and above all performing an assay for the releasing factors themselves, to discriminate physiologically active fractions from inactive ones. Once a partially purified, significantly active fraction has been obtained, it can be injected into laboratory animals for further study of its physiological effects, using well-known procedures.

The constraints on the reliability of these investigations are the sensitivity of the original assay and the purity of the isolated fractions. There is also a trade-off between the purity and the quantity of the material obtained. Huge amounts of brain tissue yielded very tiny active fractions. A significant amount of material is required for physiological studies, and therefore the more refined one's techniques of isolation are, the more money and effort must go to the initial processing of hypothalami. It was generally agreed within the field that one could do interesting, albeit not conclusive, physiological studies with only partially purified fractions; this concession in turn meant that the standards of sensitivity for an adequate initial assay were not as stringent as would be required to achieve greater isolation.

All this changes if one adopts the alternative strategy of attempting to characterize the postulated releasing factors chemically. This method cannot be a plausible strategy unless one has the resources to process tons of material and has available the skills of both a physiologist (to develop and perform the assays) and a chemist (to isolate and analyze the fractions and to synthesize analogues for further testing). It is also a risky strategy, in that it cannot succeed without first finding more sensitive assays and techniques of purification. It becomes riskier because this project makes sense only if releasing factors really are unknown substances (rather than previously characterized hormones) that nevertheless have a sufficiently well-understood chemistry to permit a straightforward chemical analysis of very small quantities of the unknown. It is still riskier in that, unlike the physiological investigations that promise a steady accumulation of results, the chemical investigation requires tremendous effort for an extended period with virtually no payoff until the hoped-for conclusion. If the effort failed to achieve a verifiable structure, there would simply be nothing to show for it.

Nevertheless, once the riskier chemical strategy is adopted by someone who has the resources required, it undercuts the validity of previous results. This state of affairs is precisely what occurred in the early 1960s. In a review paper published in 1963, Roger Guillemin set the new criteria of adequacy governing the field *as* reoriented by this new strategy and the opportunities it presented. He remarked: "These rigorous criteria contribute to take away any meaning from a great number of publications which hastily concluded that this or that substance acts only through the stimulation of the secretion of a pituitary hormone, or even that this or that protocol fits this explanation alone."[47] These conclusions were hasty, however, only from the perspective of the new program. The criteria for what counted as having a sample of TRF, LRF, or CRF were shifted.[48] It is not that some new knowledge invalidated the previous results; nothing had been *learned* to discredit them. Nor is it that someone finally noticed the unreliability of previous work. Its limitations had been well known but accepted. This work had made sense as a physiological investigation of pituitary regulation, and its results had been adequately confirmed for that purpose. But its methods, techniques, and results were virtually useless as a basis for chemical analysis. And *they were no longer adequate even for their original purpose,* since the new project had changed the operative criteria for what counted as a sample of a releasing factor. Once other laboratories were prepared to isolate these substances more thoroughly, the results obtained in earlier studies could be discounted as based upon unacceptably impure materials. It is not the impurity that is crucial, for even the new investigations produced impure samples, but rather a change in what is acceptable as a basis for serious discourse about releasing factors. How such a social change in standards could so thoroughly undermine previously accepted results that are not directly challenged by new evidence must be discussed in more detail. Before we can do this, however, we must return to our second point about the theory dominance of Heidegger's early account of science, namely, his understanding of the place of experimentation within research.

Experiment and the Construction of Microworlds

I have been arguing that scientific research is a circumspective activity, taking place against a practical background of skills, practices, and equipment (including theoretical models) rather than a systemat-

47. Guillemin 1963: 14.

48. Latour and Woolgar 1979: 116–24. Cf. Kuhn (1970a: 54–62) on the comparable ambiguities in what it is to discover oxygen, X rays, or the Leyden jar.

ic background of theory. In doing so I have already touched upon questions of experimental practice, though not thematically. We must now turn to an explicit consideration of experiment, since it presents special difficulties for a theory-dominant account of science. I noted earlier that in *Being and Time* Heidegger implicitly assigns a secondary role to experiment within the sciences. While admitting that experimental work engages us circumspectively, he excludes any place for circumspection within theoretical cognition (only the recording or communicating of theoretical results is circumspective). His discussion of science then proceeds to examine the genesis of such decontextualized theorizing while ignoring experimentation or laboratory work.[49] The omission is significant; it suggests that though experimentation is of practical importance in the actual development of scientific knowledge, it can be largely neglected when we consider the philosophical or ontological significance of the sciences. When we understand *theorizing*, we have understood what is essential about science.

Heidegger does not explain or defend his early inattentiveness to laboratory practice, but we can construct a plausible account of this omission. Three reasons seem important. The first is that Heidegger is concerned to explicate how we understand theoretical assertions, because by his account they seem initially problematic. Experimentation is easily assimilable to his account of our circumspective dealings with equipment ready-to-hand. But theoretical understanding seems initially to undermine the central themes of his hermeneutics and must be thematically addressed. This point is reinforced by the recognition that it is the comprehension of theory that is problematic, not its development or justification. The problem Heidegger is addressing is not specific to the research process; it seems to be exactly the same problem for someone not involved in science who tried to understand its claims as for someone who is actively engaged in research.

The second reason for ignoring experimentation would be the acceptance of a prevalent (albeit often implicit) view among philosophers that experimentation has no cognitive content apart from that provided by theory. This view has been powerfully expressed by Karl Popper:

> The theoretician puts certain definite questions to the experimenter, and the latter, by his experiments, tries to elicit a decisive answer to these questions and to no others. . . . But it is a mistake to suppose that

49. Heidegger 1957: 358–65, 1962: 409–15.

> the experimenter proceeds in this way 'in order to lighten the task of the theoretician', or perhaps in order to furnish the theoretician with a basis for inductive generalizations. On the contrary, the theoretician must long before have done his work, or at least what is the most important part of his work: he must have formulated his questions as sharply as possible. Thus it is he who shows the experimenter the way. But even the experimenter is not in the main engaged in making exact observations: his work, too, is largely of a theoretical kind. Theory dominates the experimental work from its initial planning up to the finishing touches in the laboratory.[50]

Such a view of the cognitive dependence of experiment upon theory,[51] coupled with Heidegger's specific concern to account for the possibility of theoretical understanding, could make it quite appropriate merely to note in passing the circumspective character of experimental practice. All that would be ontologically problematic about experimental science is the contribution made to it by theory.

The third reason for ignoring experimentation in Heidegger's early view is specific to his understanding of theory as a decontextualized cognition of the present-at-hand. Experiments, and the laboratories in which they are performed, are the local settings of a delocalized understanding. Theoretically understood, the laboratory is merely one among many possible instantiations of the theory (theories) in question. Anything specifically characteristic of the local setting or one's practical involvement in it must be irrelevant to theoretical understanding, of which the laboratory is to represent an arbitrary, interchangeable instantiation. The laboratory is analogous to the figure one constructs in a geometric demonstration. Strictly speaking, the figure is unnecessary to the proof, and nothing specific to this particular figure can be taken account of in the demonstration. Theory must, of course, hold true in the laboratory, but no more so and no differently from anywhere else. What happens in the laboratory illustrates theoretical claims and provides evidence for them, or suggests answers to questions theory has already posed, but it neither has "a life of its own" apart from theory nor makes any distinctive cognitive contribution.

This view takes the philosophically primary laboratory activities to be observing and recording. There are, to be sure, practical difficulties in designing, constructing, and running experiments, but the

50. Popper 1953: 107.

51. For some evidence that Heidegger might hold such a view about the relation between theory and experiment, see Heidegger (1952: 74, 1977: 121–22).

cognitive payoff is what one observes as the outcome of the work. The claim that observation is theory laden has caused difficulties for many philosophers of science precisely because observing was what laboratory work was supposed to be about. Experimentation was important because these observations were supposed to provide independent tests of theoretical claims. If observation was theory laden, then it would not provide an independent test. Most philosophical critics of the idea of theory ladenness have claimed that it undercuts the evidential grounding of theories. The other side of this objection is less often noted, perhaps because philosophers have been so little concerned to understand and account for the nuts and bolts of laboratory work. Traditional views of theory and experiment suggest that experiments aim to allow events of theoretical importance to be observed, and they implicitly presuppose that making these events observable does not otherwise fundamentally change them. Given such views, the claim that observation is theory laden makes laboratories initially seem superfluous. It is not so clear on such views why scientists spend so much effort acquiring and using laboratories when they are just theory by other means, and not even independent means at that.

We need an alternative understanding of the place of experiment and laboratories in scientific practice. Once again, we can turn first to neglected features of the work of Thomas Kuhn. Kuhn is widely noted for his insistence upon the importance of theory in guiding significant measurement in science. But Kuhn's own treatment of theory suggests that instrumental and experimental considerations are not entirely separable from it. Kuhn has argued that scientific "paradigms" are the fundamental locus of professional commitment in science, and he points out that he takes paradigms to be "accepted examples of actual scientific practice . . . which include law, theory, application, and instrumentation together."[52] The importance of the instrumental component of paradigms is well illustrated in his discussions of the standard test for the "goodness of air" in the work of Priestley and Lavoisier and the use of cathode-ray tubes in physics in the late nineteenth century.[53] One could just as easily cite the role of cytological techniques in chromosome genetics or of accelerators in high-energy physics. The instrumental commitments and understanding with which a theory is exemplified are an essential part of the theory as a statement about the world. Kuhn insists, for example,

52. Kuhn 1970a: 10.
53. Kuhn 1970a: 54–62.

that the tables of numbers drawn from theory and experiments "define 'reasonable agreement.' By studying them, the reader learns what can be expected of the theory. An acquaintance with the tables is part of an acquaintance with the theory itself. Without the tables, the theory would be essentially incomplete. With respect to measurement, it would be not so much untested as untestable."[54] For Kuhn, then, it is not just that theory guides the construction and interpretation of experiments; experiments and the instrumental possibilities they employ help work out what our theories actually tell us.

Ian Hacking goes considerably further than Kuhn by discussing the role of experimentation in "creating phenomena."[55] By "phenomenon" Hacking means a manifest regularity in the natural world. It is not a private sensation but a public event that commands our attention. What distinguishes a phenomenon are its clarity and reliability. Phenomena are clearly discernible, and the *circumstances* of their occurrence (although not always their causes or their most adequate description) must be well understood. The clarity of phenomena is to be contrasted to the complexity, the confused muddle, of most events in the world. In the language of information theory, phenomena have a high signal-to-noise ratio. But in this sense of the word there are comparatively few phenomena in nature. Hacking cites the preeminence of astronomy as a source of manifest regularity, suggesting that "only the planets, and more distant bodies, have the right combination of complex regularity against a background of chaos."[56] A few chemical reactions and changes of state might also be cited, along with some crystal structures, static electrical phenomena, phenotypic features of various species (especially phenomena of development), and so on. But generally nature presents us with a blooming, buzzing confusion, to adapt James's phrase to an altogether different use. An empirical science dependent upon evident regularities in nature will very quickly encounter irreducible local idiosyncrasies. The objects of such a science would be far too complex for any systematic investigation stemming from observation, however careful. We tend to say that too many factors are involved for any of them to be clearly seen, although, as Hacking notes, this can be said only in retrospect, when those "factors" have been distinguished, isolated, and analyzed. Before that there is just complexity, and the "factors" from which it supposedly results do not exist.[57]

54. Kuhn 1977: 185–86.
55. Hacking 1983: 220–32.
56. Hacking 1983: 227.
57. Hacking 1983: 226.

The advance of science has therefore largely coincided with scientists' success in creating new phenomena. Repeatable, discernible events are now commonplace in laboratories and in the technological devices that are omnipresent in our world. Laboratory technology and the technologies we employ or encounter in our daily lives have enabled us to introduce vastly greater manifest order into the world. Why is this order important? The traditional answer is that such phenomena provide the evidence for our complicated theoretical constructions. The phenomena we create in the laboratory, and the more complex occurrences they enable us to untangle, are Quine's "boundary conditions" where our theories touch on experience and are adjusted to fit. Theories tell us where to look, what to look for, and what to make of what we see. Experiment is important as a touchstone enabling us to test theory and to refine or correct it.

This answer cannot do justice to experimental practice, however. As Hacking has emphasized, experimentation in science often takes its own course, exploring areas not yet theoretically articulated. Often it is guided not by explicit theoretical hypotheses, but only by a suggestion of something worth examining and a sense of how to proceed with the exploration.[58] The principal reason for this degree of independence is that useful experimental techniques, instruments, and procedures are sufficiently difficult to develop and refine that they often play a leading role in directing experimental work. We are accustomed to crediting theoretical imagination with such a role; philosophers repeatedly emphasize that experimentation is pointless without some idea of what the experiment is supposed to reveal and why it would be important. But the error is in confining such ideas to theoretical representations. A good experimenter must have a practical grasp of the workings of her apparatus and its possibilities and limitations. This "feel" for the instruments, more a practical craft knowledge than a theoretical representation,[59] does not just tell scientists when their equipment is working properly. It also suggests possible directions for investigation—to take advantage of the capabilities of one's tools—as well as constraints on the scope and precision of the results obtainable. This feature is why I stressed the opportunism of scientific research in the preceding section. Experimentation does more than fill in the gaps left by prior theorizing or check its adequacy. It opens new domains for investigation, refines them to make

58. Hacking 1983: 149–66.

59. The importance of such craft knowledge has been discussed extensively by Ravetz (1971) and Ackermann (1985), but the emphasis on the role of tacit knowledge in science goes back at least to Fleck (1979) and Polanyi (1958).

them suitable for theoretical reflection, and provides a practical grasp of that domain as a resource. The creation of new phenomena, and scientists' practical understanding of the instrumental context of the laboratory within which this creation occurs, cannot be easily subordinated to a theory-dominant picture of how science develops.

To do justice to the creation of phenomena and the practical facility this task requires, we must redefine the place of laboratories and their apparatus within our understanding of science. The natural sciences have long been distinguished from mathematics, logic, and theology as empirical sciences. Their theories aim to refer to the world and are assessed according to the success or failure of that reference. But laboratories can no longer be regarded as merely incidental to the achievement and assessment of referential success. Reference to the world is achieved only through the mediation of the constructed world of the laboratory. And even this mediation cannot be conceived as the transfer of scientific knowledge outside the laboratory. Rather, I shall suggest (as does Bruno Latour)[60] that this involves transferring the conditions of the laboratory itself out into the world.

A laboratory on this view is not just a building or room housing the work space of scientists. Physical enclosure is a common and often useful characteristic of laboratories, but it is not essential. Fundamentally, a laboratory is a locus for the construction of phenomenal microworlds. Systems of objects are constructed under known circumstances and isolated from other influences so that they can be manipulated and kept track of. They constitute attempts to circumvent the chaotic complexity that so thoroughly limits the natural occurrence of phenomena by constructing artificially simplified "worlds." In these microworlds there exists only a limited variety of objects, whose provenance is known and whose forms of interaction are strictly constrained. Thus, for example, we might have a culture of bacteria from a specific source, a chemical solution whose components have been introduced in measured amounts from carefully manufactured and controlled stocks, or a beam of particles from an instrument constructed to produce a precisely characterizable output. The production of microworlds can be further analyzed into the "internal" isolation of laboratory components so that we know what is in the microworlds we construct (this internal component includes, for example, processes of purification and the refining of techniques or equipment) and the isolation of the experiment from any relevant external causal influences. In many cases, the object(s) to be investigated will be among the components of

60. Latour 1983.

the microworld. We know enough about the object to identify it and to characterize our interest in it, and we introduce it into a controlled microworld in order to isolate and amplify its appearance or its effects, or both. In other cases the microworld will be an entirely characterized system (at least in those respects deemed relevant to our interests). These latter cases will provide a baseline or control for comparison or will be used to introduce a well-known effect into another system.

This last activity represents a characteristic feature of laboratory practice. The point of constructing such a practically isolated microworld is to be able to *manipulate* it in specific ways. Scientists are not satisfied just to allow things to appear in isolated settings. Most scientifically interesting effects must be created by introducing new objects to a controlled setting or by manipulating the entire setting in some specific way(s). Scientists aim to create new domains for investigation by subjecting things to influences and interactions that would not occur (or would not be manifested *as phenomena*) without their intervention.

These aspects of laboratory practice reveal the error in many empiricists' Humean suspicion of causes. If we take scientists to be principally observers, then the problem of distinguishing causal efficacy from constant conjunction will perhaps seem compelling. But scientific research proceeds with the agent's presumption of causal efficacy rather than the observer's concern to discern it. Scientists deal with systems that are constructed to be causally isolated, and they deliberately introduce objects or systemic stresses that they take to be causally efficacious. Scientists make things happen in the laboratory rather than just observing what goes on around them. Thus the new empiricists such as Hacking or Cartwright, who are sensitive to the role of laboratory practice in science, differ fundamentally from many of their empiricist predecessors in their insistence upon a robust view of causation.

In addition to isolation and intervention, a third essential feature of laboratory practice is tracking what goes on in its constructed microworlds. This tracking is much more than what empiricism has emphasized under the heading "observation." Observation was supposed to be a thematic perceptual awareness of the results of an experiment, which in turn were supposed to be interpretation-free data such as dial readings, color changes, or tracks in a photograph. Tracking an experiment, however, involves monitoring the entire progress of the experiment beginning with its construction. It is a matter of seeing that things are working right rather than just seeing what the outcome is. Designing the experiment so that one can keep track of it throughout is an important part of this process. The actual activity of monitoring is less a thematic perceptual act than it is a circumspective attentiveness to the entire course of events. We are talking about the practical "sight" (and foresight) of an agent thoroughly familiar with her circumstances, a sight governed by her interests in and understanding of the proceedings. Even when something does become perceptually thematic, this monitoring is more a case of noticing something against a practically significant background than of observing and recording a datum.[61] The response may well be to adjust or redesign something rather than to record what one observes. Tracking is partly a matter of the investigator's practical skills (a "feel" for the design, attentiveness to the nuances of its operation, etc.) and partly a matter of the experiment's being constructed for visibility. Experimental design aims to make evident what is happening in the microworlds it constructs, so that its outcome will be a phenomenon in Hacking's sense.

An important, but usually overlooked, aspect of this process is the construction of a new sort of knowledge. The microworlds of scientific experiments are designed to permit detailed knowledge of their individual components. Foucault discusses extensively "the problem of the entry of the individual into the field of knowledge" but confines his discussion to the construction of knowledge of individual persons. He sees a radical transformation in what it signifies for an individual to be known.

> For a long time ordinary individuality—the everyday individuality of everybody—remained below the threshold of description. To be looked at, observed, described in detail, followed from day to day by an uninterrupted writing was a privilege. The chronicle of a man . . . formed part of the rituals of his power. The disciplinary methods reversed this relation, lowered the threshold of describable individuality and made of this description a means of control and a method of domination. It is no longer a monument for future memory, but a document for possible use. . . . The examination as the fixing, at once ritual and 'scientific', of individual differences, as the pinning down of each individual in his own particularity . . . clearly indicates the appearance of a new modality of power in which each individual receives as his status his own individuality, . . . linked by his status to the features, the measurements, the gaps, the 'marks' that characterize him and make him a 'case'.[62]

As Foucault insists, this knowledge of individuals results not just from observation, but from "the whole apparatus of writing that accom-

61. Hacking 1983: 167–85.
62. Foucault 1977: 191–92.

panied it. . . . These small techniques of notation, of registration, of constituting files, of arranging facts in columns and tables that are so familiar to us now, were of decisive importance in the epistemological 'thaw' of the sciences of the individual."[63] Tracking the components of the laboratory's microworlds is a matter of sorting, coding, filing, and recording the identity, location, and disposition of every one of them.

Foucault confines his concern to the practices that constitute a knowledge of individual human beings. But a similar development takes place in the laboratory, using parallel techniques. The phenomena created in the laboratory are not recorded as a monument to their intrinsic importance. On the contrary, they are important only because they can be tracked and recorded throughout their various manifestations. They are important because they are the objects of a detailed individual knowledge, which thereby subjects them to new forms of manipulation and control. The significance of this situation for scientific practice and the knowledge it embodies is illustrated by a thought experiment conducted by Bruno Latour and Steve Woolgar.

> Imagine that the following experiment was carried out by our observer. Entering the laboratory at night, he opens one of the large refrigerators. As we know, each sample on the racks corresponds to one stage of the purification process and is labelled with a long code number which refers back to the protocol books. Taking each sample in turn, the observer peels off the labels, throws them away and returns the naked samples to the refrigerator. Next morning, he would doubtless witness scenes of extreme confusion. No one would be able to tell which sample was which. It would take up to five, ten, and even fifteen years (the time it took to label the samples) to replace the labels—unless, of course, chemistry techniques had advanced in the mean time. As we stated earlier, any sample might equally well be any other. In other words, the disorder, or more precisely the entropy, of the laboratory would have increased: anything could be said about each and every sample.[64]

The maintenance of ordered microworlds in which the order is isolated, monitored, and preserved throughout is the prerequisite for scientists' creation of phenomena. Beyond having a (theoretical) idea of where to look and what to look for, one needs to create the field within which what one is looking for can become clearly visible. This is what laboratories are for in scientific research.

I have taken the term "microworld" from Marvin Minsky and

63. Foucault 1977: 190–91.
64. Latour and Woolgar 1979: 244.

Seymour Papert.[65] They use it to refer to isolated fictional domains whose objects and possible events can be represented in a computer program as the background that enables the computer to simulate understanding and intelligent behavior within that restricted domain.[66] Of course, in my usage it refers not to fictional domains but to isolated systems actually created in the laboratory. But the connection to Minsky's and Papert's conception of fictional microworlds is important, because it suggests the possibility that laboratory microworlds might be partial or approximate realizations of the fictional systems depicted in theoretical models. On Nancy Cartwright's view, discussed earlier, the content of theories is expressed in the fictionalized models or simulacra within which the (highly idealized) implications of the theory can be calculated. I now want to suggest that such theoretical modeling finds its experimental counterpart in the construction of microworlds that more closely resemble the simulated microworlds of the theoretical simulacra. The point is not that theoretical models are actually realized in the laboratory, since often the idealizing assumptions of the modeling are in principle unrealizable or present unmanageable practical difficulties.[67] Rather, scientists create experimental analogues to such models, which like them are isolated systems, which often attempt to distinguish more and isolate more effectively objects, events, or processes to which theories might refer, but which would not otherwise appear unmasked from more complicated, unanalyzed situations in the world at large. Experimental microworlds are thus generally designed to be amenable to theoretical treatment.

It should not be supposed, however, that the relations between theoretical models and laboratory microworlds are unidirectional. Sometimes, to be sure, experiments are designed with specific theoretical models in mind. But experimental situations are also commonly constructed that may aim to be amenable to theoretical modeling but for which no such model actually has been formulated. Imaginative theoretical development or extensive calculation may be required to develop such a model. Yet the experiment may be constructed with the *possible* parallel formulation of such a theoretical model in mind. This state of affairs seems to be what happened in the

65. Minsky and Papert 1973: 95.

66. The prototype for this conception of a computer-represented microworld was Terry Winograd's program SHRDLU, discussed in Winograd (1973).

67. Cartwright (1983: 156) reminds us that even when theory provides very close analogues to experimentally contrived situations, this does not justify treating the model realistically, even in the case of the analogue.

case cited earlier of Raymond Davis's attempt to measure the solar neutrino flux. Davis's experimental design reflected no preexisting theoretical treatment of the production of solar neutrinos. But it was designed to make possible a theoretical modeling of his laboratory microworld, and indeed the experimental design and the theoretical modeling were developed hand in hand.[68]

Robert Ackermann is thus undoubtedly correct in suggesting a dialectical relationship between the development of theory and experimental data.[69] But Ackermann does not go quite far enough, because he seems to suggest that the interaction is primarily between the data texts that result from the experimental deployment of readable instruments and an undifferentiated notion of theory. Had he discussed the model structure of theories and the microworlds from which his "data domains" are generated, he could have given a finer structure to the dialectic he rightly sees taking place. Theories are adjusted not just to respond to more refined data, but to deal with newly developed experimental situations of which the data are only the most prominent representative. Experiments respond not just to theoretical ideas, but to the various models in which those theories are worked out. The experimental situation must in turn be made describable in terms the equations embodied in the theoretical models can attach to. As Cartwright has emphasized, this requirement is a distinct level of concern that presents its own difficulties[70] and could well introduce considerations into the experimental design. With these considerations in mind, the best philosophical account of the dialectic of theory and experiment may still, twenty years later, be the third chapter of Kuhn's *Structure of Scientific Revolutions* (keeping in mind his emphasis on the paradigmatic structure of theories).

Let us return to the example of Davis's measurement of solar neutrinos to help clarify the distinction between laboratories and the microworlds they contain. Davis's experiment constructed a single microworld, consisting of a measured volume of perchloroethylene (containing the Cl-37 isotope of chlorine) and a continuous stream of particles passing through it. His "laboratory," however, included the sun, which was the source of much of his particle stream, several miles of the earth's crust, which effectively isolated the desired particle stream from the background of cosmic radiation, and the detection devices outside the tank to extract and count the argon-37 atoms

68. Pinch 1980.
69. Ackermann 1985, chap. 4.
70. Cartwright 1983, essay 7.

expected to result from neutrino-chlorine interactions. Laboratories, then, need not be bounded by walls. They extend outward to incorporate features of the natural world within the artificial worlds they construct. Like the workshop as Heidegger discusses it in *Being and Time*, the laboratory is not a physical space between four walls but a context of equipment functioning together, which even incorporates nature among that equipment.[71]

In this sense the laboratory, like the clinic, the asylum, the school, the factory, and the prison, serves as one of the "blocks" within which, according to Foucault, a "micro-physics of power" is developed and from which that power extends to invest the surrounding world. Foucault, admittedly, distinguishes the exercise of power over persons from the development and deployment of our capacities to deal with things.

> As far as this power is concerned, it is first necessary to distinguish that which is exerted over things and gives the ability to modify, use, consume or destroy them—a power which stems from aptitudes directly inherent in the body or relayed by external instruments. Let us say here it is a question of "capacity." On the other hand, . . . let us not deceive ourselves; if we speak of the structures or the mechanisms of power, it is only insofar as we suppose that certain persons exercise power over others.[72]

But though they can be distinguished, powers and capacities cannot be separated in their actual deployment. Foucault notes:

> There are also "blocks" in which the adjustment of abilities, the resources of communication, and power relations constitute regulated and concerted systems. . . . These blocks, in which the putting into operation of technical capacities, the game of communications, and the relationships of power are adjusted to one another according to considered formulae, constitute what one might call, enlarging a little the sense of the word, disciplines.[73]

In chapters 6 and 7 I shall argue that this intertwining of powers and capacities is even more fundamental than Foucault thinks; an essential aspect of the ways "certain actions modify others"—that is, in which power functions—is by transforming the world within which action is situated. This transformation is the most fundamental way

71. Heidegger 1957: 68–72, 1962: 97–102.
72. Foucault 1982: 217.
73. Foucault 1982: 218–19.

power (even in Foucault's restricted use of the term) operates within and extends outward from the laboratory. But the techniques of surveillance and examination (and even of "normalization," albeit employing different means) that Foucault sees at the core of the disciplinary construction of the modern subject have their analogues in the construction and manipulation of microworlds within the laboratory. And as we shall see, even the diffusion of knowledge and power from the laboratory follows a pattern Foucault has already discerned in the disciplinary blocks that have most concerned him.[74]

Before this process can be considered in detail, however, we must draw together the preceding discussions into an argument for the claim that scientific knowledge is fundamentally local knowledge, embodied in practices that are not fully abstractable into theories and context-free rules for their application. In Heidegger's terms, science must be understood as a concerned dwelling in the midst of a work-world ready-to-hand, rather than a decontextualized cognition of isolated things. Many features of scientific research as I have been describing it converge in support of this claim. We have seen that the empirical character of scientific knowledge is established only through the use of a local configuration of equipment in the laboratory. The instruments, and the microworlds established by using them, are the proximate referents of scientific claims. Furthermore, scientists' knowledge depends upon their craft knowledge of the workings of these devices. Knowing one's way about the local setting of the laboratory is an indispensable part of scientists' achievement.[75] Yet this knowledge cannot readily be conveyed without drawing upon such accumulated local experience.[76] As Ackermann notes: "An instrument can be named or illustrated in an article, or a modification in technique noted. These sorts of references do occur. But the feel of an instrument and an explanation of its working cannot easily be accomplished verbally. Theory and data can be described, but one must learn how to operate an instrument."[77] Associated with the local laboratory context, then, are the skills researchers gradually build up in developing and adapting the equipment to locally determined needs and purposes. The importance of such skills is attested in Ludwik Fleck's discussion of the Wassermann reaction as an established scientific fact. Fleck claims that

74. Foucault 1977: 209–28.
75. Hacking 1983: 227–30; Ravetz 1971: 75–83, 88–103, 173–75.
76. Knorr-Cetina 1981: 127–30.
77. Ackermann 1985: 132.

> Wassermann's reports about his reaction contain only the description of the relation between syphilis and a property of the blood. But this is not the most important element. What is crucial is the *experience* acquired by him, by his pupils and in turn by theirs, in the practical application and effectiveness of serology. Without this *experience* both the Wassermann reaction and many other serological methods *would not have become reproducible and practical.* Such a *state of experience* became general only slowly and had to be practically acquired by each initiated individual. . . . Even today, anybody performing the Wassermann reaction on his own must first have acquired comprehensive experience before he can obtain reliable results.[78]

Although the experimental manipulations in which phenomena like the Wassermann reaction are created are the most obvious place in which local, experiential skills are essential to science, even theoretical and mathematical tools have an experiential basis. Learning how to attach these techniques and models to particular situations, and to avoid the various problems and pitfalls that arise there, requires tacit craft skills as much as any experimental procedure.[79]

Even the local community of co-workers is a significant factor in the development and reproduction of knowledge. We have seen this effect already in our discussion of the opportunism of scientific research. Who is available (i.e., the local configuration of technical skills and other tacit know how) affects what counts as an opportunity worth pursuing. But even this local ensemble cannot be understood just as a collection of individually skilled persons. Scientific research capabilities often require specific patterns of teamwork, which cannot be acquired without actually working together for some period of time using the same equipment and local procedures. In the case of the Wassermann reaction, Fleck points out that "change in personnel often produces a disturbance in the progress of the reaction, even if the new member of the team had worked well with other associates. This explains the poor results obtained even by excellent research workers at the . . . Wassermann conferences held under the auspices of the League of Nations."[80] Research skills do not immediately transfer outside their physical setting and local community; one must first learn one's way about in the new context and adapt one's prior knowledge to it.

78. Fleck 1979: 96.
79. Ravetz 1971: 88–101.
80. Fleck 1979: 97.

Part of the reason for this difficulty is that scientific research is beset with local idiosyncrasies. The precise problems encountered in any given laboratory working on a particular project are likely to be unique. As one biochemist noted in an interview,

> Basic work is usually done . . . on something similar, but not the same. You know, if it's done on what I am interested in, then it's not worth doing again. So usually it's done on something similar. . . . And see, I think you almost always have to adapt (a method) in some way. Sure, occasionally you find something that just fits in perfectly to solve a problem—but I'd say that's the exception rather than the rule.[81]

This is also true in the case of long-standardized procedures, such as the Wassermann reaction, as each laboratory develops its own variations on the standardized scheme.[82] Surprisingly enough, it turns out to be the case even when the original aim of the work is to replicate someone else's (apparent) achievement. It is not even clear that an experiment could be replicated exactly, but in practice scientists deliberately develop their own modifications and alternative approaches to what initially seems the same procedure.[83] They do so partly because they have their own preferred techniques, approaches, and standards, because they often believe they can do it "better," because they must adapt to local conditions, and because they do not want their work to appear superfluous. But they do so also because much of the practical procedure, especially the mundane adjustments required to make things work reliably, cannot be reported in published accounts of experimental work.

> Many laboratory problems which are left unthematized in the papers can be associated with the development of scientific "know-how" which . . . is a *local* (or even personal) practical knowledge of how to make things work. For every bit of published "method," there seems to be a bit of unpublished know-how which not only reconstructs the recipe sequence of steps in the paper into that of feasible doings within the situational logic of laboratory action, but also provides routines for diagnosing and coping with many unspecified problems.[84]

These are not just details that any competent practitioner could fill in if she wanted. Knorr cites interviews with scientists who point out that

81. Knorr-Cetina 1981: 38.
82. Fleck 1979: 53.
83. Collins 1982.
84. Knorr-Cetina 1981: 128.

informed, thoughtful readers cannot reconstruct experimental procedures and that even the coauthors of a paper may not understand or be able to reproduce the procedures described.[85] The local, practical understanding of how to make things work in the laboratory cannot be readily transferred outside without transferring the people involved and even relevant features of the physical setting itself.[86]

These skills and procedures are aimed toward practical success (getting things to work reliably) rather than toward systematic understanding or truth.[87] This emphasis is empirically evident when difficulties or "pitfalls" in one's instruments or procedures are circumvented rather than systematically explored and understood.[88] Yet it is only within such local, intensely pragmatic contexts that the objects of most scientific research exist. I must insist upon this point. The phenomena that provide scientists with both material for investigation and the data against which their theoretical models are developed and refined do not exist outside such pragmatically constructed situations. The instruments that enable the phenomena to appear do not exist outside laboratories or other places constructed to duplicate laboratory effects. The phenomena themselves are usually manufactured, isolated, and manipulated within the laboratory. The clearly structured data that result (when the work succeeds) always reflect the local, contingent, idiosyncratic circumstances of their production. Knorr appropriately concludes:

> All of the source-materials have been specially grown and selectively bred. Most of the substances and chemicals are purified and have been obtained from the industry which serves science or from other laboratories. But whether bought or prepared by the scientists themselves, these substances are no less the product of human effort than the measurement devices or the papers in the desks. It would seem, then, that nature is not to be found in the laboratory, unless it is defined from the beginning to be the product of scientific work.[89]

Standardization or Decontextualization?

But if the irreducibly local character of scientific knowledge is so evident, why is it so thoroughly at odds with the universalism that has

85. Knorr-Cetina 1981: 128–29.
86. I have in mind the transfer of locally developed instruments and materials. The need for this is often not apparent after initial work has been done, because both instruments and materials are often then manufactured and sold in standardized forms. This will be discussed further below.
87. Knorr-Cetina 1981: 4.
88. Ravetz 1971: 97–101.
89. Knorr-Cetina 1981: 4.

traditionally dominated our understanding of science? No doubt the latter is partly an artifact of the theory bias of many of the philosophers and scientists who have helped shape the image of science within our culture. But ultimately this answer is unsatisfactory. For scientific knowledge does often appear shorn of contextual reference, and the ability to extend scientific capabilities outside the laboratory has been a hallmark of modern science and has been especially prominent in shaping its cultural image. We cannot regard the foregoing account of science as fundamentally local knowledge to be well established unless we can adequately account for the apparently widespread evidence for its transcendence of the local.

Two features of science seem especially persuasive initially as objections to describing it as local, pragmatic know-how. These features are its technological applicability outside the laboratory and its decontextualized expression and transmission in the scientific literature. In the latter case, for example, Latour and Woolgar point out that there is typically in scientific research "a point of stabilisation, when a statement rids itself of all determinants of place and time and of all reference to its producers and the production process."[90] This point has been seized upon by Hubert Dreyfus to defend Heidegger's original account of science as a decontextualizing activity that sheds its practical involvements with local contexts of equipment ready-to-hand. Dreyfus goes further to argue that this characteristic forces a wedge between the natural sciences, for which a Quinean theoretical hermeneutics of translation provides a cogent interpretation, and the study of human beings, which can never escape the local contingency and idiosyncrasy of its objects and methods.[91] Dreyfus argues that all sciences involve practices of interpretation, for which Heidegger's account of our practical involvement with the ready-to-hand is appropriate. But he insists that although even the natural sciences *presuppose* such practical achievement, their results, unlike those from the study of human beings, attain independence of any reference to the particular circumstances of that achievement. This differential independence from the circumstances of their production also neatly explains the predictive power and technological applicability of the natural sciences in contrast to the relative lack of predictive success in other disciplines. Natural scientific knowledge can be formulated into universal laws from which indefinitely many local instantiations can be deduced, making possible its widespread application. On Dreyfus's

90. Latour and Woolgar 1979: 175–76.
91. Dreyfus 1980, 1984.

account, then, hypothetico-deductive accounts of the transmission and application of scientific knowledge can be supplemented by sociological studies of the contingent circumstances of its production, but these contingencies are incidental to the knowledge itself. We seem to have come full circle to something like the traditional account I initially rejected.

There is, however, an alternative conception of what is going on here, which I shall argue does more justice to the phenomena than does Dreyfus's return to the early Heidegger. I have been arguing that the laboratory is a field of equipment (both physical and intellectual) for creating, manipulating, and interpreting phenomena. What Dreyfus takes to be a "theoretical" decontextualization of those phenomena is more akin to the transformation of a tool originally designed for a highly specific task within a particular context into a more general-purpose item of equipment. When hammers are produced in standard sizes and forms (even "all-purpose" ones) rather than custom designed, or when clothes are manufactured in standard sizes, they lose any built-in reference to specific persons, tasks, or situations, but they do not cease to be ready-to-hand. In Heidegger's terms, their functionality is "averaged" out.

> The work produced refers not only to the "towards-which" of its usability and the "whereof" of which it consists: under simple craft conditions it also has an assignment to the person who is to use it or wear it. . . . Even when goods are produced by the dozen, this constitutive assignment is by no means lacking; it is merely indefinite, and points to the random, the average.[92]

Following Ravetz, I shall call such averaging the "standardization" of scientific problems, tools, procedures, and results.[93] It involves both transforming the things themselves to make them applicable outside their original setting and developing more exoteric interpretations that make them accessible to the nonspecialist.

The standardization of the products and tools of laboratory work does indeed remove most references to the particular local contingencies of their production, but only by making these references indefinite. It does not abolish references to the field of activities and locales within which things of this kind acquire intelligibility. And like all general-purpose tools, laboratory work always reacquires its particular intellibility by being readapted to the particular circumstances of

92. Heidegger 1957: 70–71, 1962: 100.
93. Ravetz 1971: 196–202.

its various uses. This effect we have already seen, for example, in Fleck's account of the adaptation of the now standardized scheme of the Wassermann reaction to the particular circumstances of a given laboratory.[94]

Indeed, evidence of such standardization is prevalent in recent studies of scientific research in context. Ravetz, from whom I adopted the term, thematizes this process. He discusses extensively the methods by which problems, tools, facts, and the objects of investigation are transformed into versions that can be extended beyond their original domain, though not without some loss of content and clarity.[95] The most important feature of this process is the avoidance of pitfalls. As Ravetz notes:

> Only when the tool will be involved in the production of a crucial or delicate piece of evidence is it necessary for its user to have a full appreciation of its possibilities and pitfalls. If it is applied in a routine fashion in the production of data, then a simple, rough-and-ready understanding is sufficient. What is desired in such contexts is a standardized version of the original tool, robust rather than refined in its design.[96]

A good example of this standardization is the evolution of gene cloning over the past decade. When Cohen, Berg, and their colleagues first recombined genetic fragments in a plasmid and inserted this into a bacterium it was a very esoteric project, an object of research by specialists rather than a tool to be employed routinely for other purposes. One needed a supply of Cohen's specially constructed plasmid even to get started. Now cloning is a standard technique used in laboratories in many subfields.[97] A variety of plasmids are available from commercial laboratory supply houses, and there are now "cookbook" procedures for recombining and cloning them that make a specialist's understanding of the technique otiose for most purposes.

Standardization thus makes a technique more robust and more exoteric. By building past experience into the procedures and equipment, standardized procedures are also made less sensitive and more stable. This process is the correlate to their being extendable beyond the domain of specialized use. Fleck describes the extent to which this occurred in the development of the Wassermann reaction:

94. Fleck 1979: 53; see text at note 82 above.
95. Ravetz 1971: 195–208.
96. Ravetz 1971: 196.
97. It has even found its way outside biology proper into physical anthropology, as exemplified by Neiswanger (1985).

> During the initial experiments it produced barely 15–20 percent positive results in cases of confirmed syphilis. How could it then increase to the 79–90 percent found in later statistics? . . . The moment when this decisive turn occurred cannot be accurately determined. . . . even the principal actors themselves can say no more than that the technique had first to be worked out. Sometimes Citron is credited with having brought about the turning point through his introducing increased serum dosage. Wassermann and his co-workers originally used 0.1 cc. of patient serum, but Citron recommended 0.2 cc. Yet today even 0.04 cc. of patient serum is ample, provided all the reagents are mutually adjusted with precision. Fundamentally it is this very reagent-adjustment, coupled with learning how to read the results, that made the Wassermann reaction useful.[98]

The serologist Edmund Weil wrote at the time that "the technical development of the Wassermann reaction . . . proceeded in the direction of making the reaction less and less sensitive to obtain a clinically useful test for syphilis."[99]

Latour and Woolgar found a similar process, with a more dramatic transition from the esoteric to the exoteric, in the investigation of thyrotropin releasing hormone (TRH) in the late 1960s. Before 1969, TRH could be identified and confidently discussed only within a few specially equipped laboratories whose workers had devoted a decade to isolating and characterizing it. Only very minute, partially purified samples of TRH existed. The name "TRH" referred to whatever in those samples caused the differences between the samples and control substances in specific physiological assays. As Latour and Woolgar describe the situation, "to find out what TRH was before this date would have entailed a laborious search through a complex mesh of forty-one papers, full of contradictory statements, partial interpretations, and half-baked chemistry."[100] What made the difference after 1969 was the ability to synthesize comparatively large quantities of virtually pure TRH reliably and automatically (using the automatic peptide synthesizer). TRH was manufactured in a form that could circulate outside the original small circle of laboratories, with their fractionating columns and sensitive physiological assays.[101] And yet TRH, in its new formulation as synthetic pyro-glu-his-pro-amide, had not been stripped of all reference to laboratories. This reference had been generalized and now encompassed a much more indefinite vari-

98. Fleck 1979: 72.
99. Quoted in Fleck 1979: 73.
100. Latour and Woolgar 1979: 148.
101. Latour and Woolgar 1979: 148–49.

ety of uses; but its existence and above all its use were still intelligible only within fairly specific equipmental contexts. If laboratories and the activities associated with them were to be destroyed, synthetic pyro-glu-his-pro-amide would lose its significance.

Virtually every kind of scientific achievement can be standardized and can thereby circulate outside its original more specific context. We have now seen the obvious case of instruments and procedures. (Of course, in science instruments are themselves often a reified standardization of a procedure, removing both the need for specialized skill to perform it and some of the possibilities for procedural error or procedural innovation; the automatic peptide synthesizer mentioned above is a good example of this.) But facts and theories also have their standardized versions. Sometimes standardized facts are just the correlates of standardized procedures. The correlation between syphilis and a complicated transformation of the blood is a fact that was at first manifested in a very limited context and gradually, through repetition and practical refinement of the procedure, was extended into new domains. But this fact can never be fully disentangled from the procedures and equipment both for establishing the existence of the amboceptor in the blood and for independently identifying and diagnosing a disease entity, "syphilis."

We can come to a similar understanding of the evolution of theories. The model Hamiltonians Cartwright discusses are simply the standardized formulations of quantum mechanics.[102] The series of refinements in which theories move from their original formulation in the journal literature to the textbook version that is cleaned up and adapted for a wide variety of envisioned uses is the institutionalized setting for this process.[103] It is useful in this respect to think of theories as strategies for dealing with various phenomena instead of systems of statements, as policies rather than creeds.[104] To formulate a description of a phenomenon is to opt for a particular way of dealing with it, and this applies with particular force to theoretical descriptions, if we keep in mind the role of models or exemplars in our understanding of theories. Theories are always adapted to their possible use in exploring and manipulating the phenomena they are used to describe and explain. Indeed, Knorr observes that theories in use tend to disappear from thematic concern, in a way reminiscent of

102. Cartwright 1983: 136–39.

103. See Fleck 1979: 118–25; Ravetz 1971: 191–208.

104. I owe this last formulation to the physicist J. J. Thomson, quoted in Knorr-Cetina (1981: 4).

Heidegger's account of the phenomenal "withdrawal" of equipment.[105] "Theories adopt a peculiarly 'atheoretical' character in the laboratory. They hide behind partial interpretation of 'what happens' and 'what is the case,' and disguise themselves as temporary answers to 'how-to-make sense-of-it' questions. What makes laboratory theories so atheoretical is the lack of any divorce from instrumental manipulation."[106] Theories, like other equipment, recede from thematic attention in order to highlight the things they make accessible to us. They themselves become the objects of attention only when we need better equipment to deal with the problems at hand. Science exhibits a rapidly oscillating dialectic between concern with theories and concern with the phenomena they describe primarily because it constantly works at the limits of its equipment's capabilities and aims to extend those limits. Those problems science can readily solve are no longer of scientific interest, except insofar as they can be incorporated into equipment that can carry research into new domains.[107] Standardization is the process of adapting the work of the vanguard of research into material usable in further investigation long after the vanguard has left that work behind.[108]

The importance of describing this process as the standardization of something ready-to-hand rather than as a "theoretical" decontextualization becomes clear when we realize that such averaging out of our understanding always involves some loss of content. Cartwright emphasizes this point throughout her account by insisting that the explanatory power of standardized theoretical models in physics is purchased at the expense of phenomenological truth.[109] Fleck recognized that in experimental work making a procedure more generally applicable and exoteric also makes it less sensitive and discriminating.[110] Ravetz attempts a more general account of this trade-off, arguing that the decay of information that results from standardizing and generalizing facts, tools, and objects of inquiry can be likened to the physical process of entropy increase in an ordered system.[111] In each case, however, we are reminded that the removal of traces of their contextual origin from scientific results reflects a pragmatic

105. Heidegger 1957: 70, 1962: 99.
106. Knorr-Cetina 1981: 4.
107. See Ravetz 1971, part 2, passim.
108. On the role of the vanguard, see Fleck (1979: 124).
109. Cartwright 1983, essays 7–8.
110. Fleck 1979: 73.
111. Ravetz 1971: 200–208.

choice within a larger field of practical involvements, a balancing of gains and losses, rather than a move from local practical involvement to a universal "theoretical" stance.

I hope I have now satisfactorily dismissed Dreyfus's objection to the claim that science produces fundamentally local knowledge. The local reference of scientific theories, facts, and other equipment can be made indefinite, but it is never stripped away entirely to reveal isolated things present-at-hand. But there was the related objection that our account could not explain the technological applicability of scientific knowledge. I shall argue that such application involves a process of extension similar to standardization as I have been describing it. But technological applications of science also bring to light a feature of this process that occurs within science itself, though less evidently. To extend the know-how scientists develop beyond its local context, one must not only refine and adapt the procedures, strategies, and equipment themselves but must also partially reconstruct the situation within which the know-how is to be applied. Science sometimes "works" only if we change the world to suit it. These changes are less evident within science, because much of what is involved comprises general features of the construction of laboratories. They are also partially concealed in the case of technological extensions by the massive general effort within industrialized societies to make widely applicable the standardized units of measurement, the common techniques of observing, counting, measuring, and monitoring things, and the purified substances and specialized equipment that give scientists within laboratories rigorous control over the microworlds they construct. These practices cannot always be tied to particular applications of science. But as Bruno Latour has recently suggested, for example,

> most of the work done in a laboratory would stay there forever if the principal physical constants could not be made constant everywhere else. Time, weight, length, wavelength, etc., are extended to ever more localities and in ever greater degrees of precision. Then and only then, laboratory experiments can be brought to bear on problems occurring in factories, the tool industry, economics or hospitals.[112]

Nor is this effort trivial. J. S. Hunter has recently estimated that the effort required to maintain the constancy of such physical constants is three times that directly involved in science and technology.[113] If one includes among the conditions of the application of scientific knowl-

112. Latour 1983: 166–67.
113. Hunter 1980.

edge such things as the development of the chemical industry to manufacture the pure substances that chemical know-how presupposes, the vast transformation in farming practices that has permitted the extensive application of agricultural science, or the development of machine tools capable of fabricating metals to exact tolerances, the effort involved begins to appear staggering. Latour's description of these developments as "the transformation of society into a vast laboratory"[114] begins to seem less far-fetched. It parallels Foucault's account of how forms of power developed within the confines of prisons or hospitals came to circulate in the society surrounding them. Foucault notes in *Discipline and Punish* that "while, on the one hand, the disciplinary establishments [e.g., hospitals, prisons, and schools] increase, their mechanisms have a certain tendency to become 'de-institutionalized', to emerge from the closed fortresses in which they once functioned and to circulate in a 'free' state; the massive, compact disciplines are broken down into flexible methods of control, which may be transferred and adapted."[115] Just as Foucault finds it not "surprising that prisons resemble factories, schools, barracks, hospitals, which all resemble prisons,"[116] Latour thinks we should not be astonished to find that laboratory know-how tells us something about the world around us. We have helped *make* the one relevant to the other.

The description of the extension of scientific knowledge outside the laboratory as a "translation" of local practices to adapt to new local situations (themselves altered to ease the transition) has, I think, begun to be plausible. The claim is not that scientific knowledge has no universality, but rather that what universality it has is an achievement always rooted in local know-how within the specially constructed laboratory setting. The empirical character of scientific knowledge is the result of an irreducibly local construction of empirical reference rather than the discovery of abstract, universal laws that can be instantiated in any local situation. Too much of scientific knowledge involves preparing the situation to make laws applicable to it and learning how to describe it in terms to which laws can attach. Such preparation and description always constitute a form of local knowledge.

The Social Context of Justification

We have now seen that scientific research is an involved, practical activity rooted in the skillful grasp of specially constructed local situa-

114. Latour 1983: 166.
115. Foucault 1977: 211.
116. Foucault 1977: 228.

tions, typically laboratories. But it must also be understood to be socially situated. The intelligibility of the laboratory itself, of the manipulations scientists perform within it, and of the maneuvers they make in reporting their results rests upon the anticipated response of others. Science is an activity belonging to *Dasein,* a behaviorally self-adjudicatingly social way of interpreting oneself and things. Any particular scientific project thus acquires significance only against the understood background of what one does in this sort of research and what might actually be said in response to it. Scientific claims are formulated and defended with respect to the concerns and possible objections of an indefinitely extended group of scientists to whom they are addressed. There is little concern to argue for them outside this specific social context. Scientific claims are thus established within a rhetorical space rather than a logical space; scientific arguments settle for rational persuasion of peers instead of context-independent truth. Or rather, what it means for scientists to argue for the truth of their claims is to attempt rational persuasion of their peers.

Hilary Putnam has recently taken an important step toward recognizing this distinction. He claims that

> someone's telling us that they want us to know the truth tells us really *nothing* as long as we have no idea what standards of rational acceptability the person adheres to: what they consider a rational way to pursue an inquiry, what their standards of objectivity are, when they consider it rational to terminate an inquiry, what grounds they will regard as providing good reason for accepting one verdict or another on whatever sort of question they may be interested in.[117]

In science, the standards of rational acceptability are not individual but social. They are embodied in the institutions within which research is refereed for funding or publication, cited and reviewed, and reformulated for inclusion in standard texts and reference works. But underlying these institutionalized procedures are the perhaps even more influential informal networks within which techniques and results are evaluated in the context of ongoing research. Scientists repeatedly ask themselves and their co-workers what they can rely upon or adapt to their own use, what they must take account of, what they can ignore or must do for themselves. The standards and arguments built up in these informal evaluations are constantly shifting to take

117. Putnam 1981: 129.

account of new results and the possibilities they open up, but they form the most basic level at which scientific work is assessed.

These practical standards of rational acceptability are a clear example of the behavioral self-adjudication that Heidegger describes as *Dasein*. These standards determine what counts as legitimate scientific work that must be taken into account, and they provide a criterion for determining who belongs to the community of research workers and its various subcommunities. At the same time, however, these standards derive their authority from their use by those persons who already count as doing science and as belonging to a particular field of work. They must be taken into account, that is, precisely because they represent "what one does" in that particular field. Anyone who wishes to be taken seriously in that line of research must address those standards, and anyone whose work is judged to do so will normally be taken seriously.

We are not talking primarily about abstract or formal standards of rationality here. The important issues are usually substantive: what counts as competently constructed data, what methods are reliable, which citations are controversial and which are not, what claims require argument, which alternative interpretations must be foreclosed, and so forth. Ad hominem argument plays a perhaps surprisingly significant role in the informal discussion that shapes the direction of research; the adequacy of data and procedures is often assessed on the basis of the credibility of the investigator.[118] The other side of this situation is that investigators tend to be seriously concerned to maintain their own credibility. And this task involves adhering to what they think are the standards others will bring to bear in evaluating any particular claim. As a result, what can permissibly be concluded in any scientific report will be the subject of intense negotiation, first between coauthors, later perhaps between authors and referees. Much of the work of such negotiation is accomplished at the original stage of writing a paper, however, as the authors assess how readers in the field are likely to respond and how the paper and the research it is based on can be adjusted to head off anticipated questions or criticisms.[119] Scientific experiments are designed, and papers written, in response to the specifically anticipated concerns of other scientists rather than some context-free standard of rationality. Nor is this ap-

118. Latour and Woolgar 1979: 155, 163–65, 187–230.

119. Knorr-Cetina (1981: 94–133) has extensively analyzed the process of negotiation reflected in the construction of a single paper. See also Latour and Woolgar 1979: 154–67.

proach inappropriate. Scientific research proximally aims to contribute to an ongoing shared project of investigation, not to a cumulative collection of confirmed or corroborated statements. The standards scientists employ, then, are what they take to be the practical standards of their fellow participants. Note that this understanding involves two distinct claims: the standards are *practical*, based upon the perceived needs of ongoing research activities, and they are *situated* within localized social networks.

We should not, however, make the mistake of taking these community networks as composed of monolithic groups, with a consensus about fundamental issues in their work. I have tried to emphasize throughout the book the important difference between shared practices, equipment, and issues and shared beliefs or values. There is substantial disagreement within scientific fields, including differences in the evaluation of particular achievements and even in the standards applied to such evaluation. But there are limits upon the range of legitimate disagreement. These limits play the important role of defining the self-adjudicating group of practitioners; in turn, scientists take account of the range of opinion within those limits when they address their work to the group and not some mythical community consensus. Even here, differences emerge. One must take account of objections to one's work raised from within the limits of legitimate community disagreement, but there may be differences among research groups in their respect for divergent opinions, and this attitude in turn will be affected by one's respect for the particular researchers one expects to hold those opinions.

All of this is complicated by the fact that standards of evaluation in science are dynamic. New results, techniques, and interpretations compel revisions in what can be taken to be adequate data, reasonable analogies, tolerable misfit between theory and data, and so forth. Such new developments are themselves going through the process of criticism from which settled scientific claims emerge and are therefore controversial. Yet this unsettled, controversial "forefront" of research also exercises a powerful influence upon the critical process itself. William Macomber once remarked:

> It is the context [of scientific knowledge as a whole] and not the individual discovery which counts; the discovery derives its significance from the contribution which it makes to the context. This context has constantly to be reorganized and renewed. If we search for it, we cannot find it, for it is constantly expanding, shifting, being modified. The source of the significance of individual discoveries and the basis of their

> validity, it nevertheless always eludes our grasp. Yet it is a context with which every working scientist is familiar; among scientists it is what "everybody knows."[120]

Yet what "everybody knows" cannot be identified with any stage of the published literature in science, since it often runs ahead of publication.[121]

This observation also raises a deeper point about the criteria of rational acceptability in science. Scientific discourse takes place at many levels, from informal conversation to formal presentations or publications and eventually to retrospective review and incorporation into reference works. The standards applied to a statement will depend crucially upon both the context in which it was made and the context of evaluation. In informal conversation, reference can and should be made to the most up-to-date, unpublished results, but one can also be freer with one's assessments. Artifacts are thus more readily given credence, and serious work is more readily dismissed in this context, yet it helps form the "cutting edge" of scientific advance. Research not yet authenticated by the social processes of criticism and standardization can help shape the subsequent application of those same processes. The reduced constraints upon acceptable statements in this context reflect their ephemeral character. They involve tentative judgments open to almost immediate revision. The journal literature reflects both more rigorous and formal standards of accountability and a willingness to admit dubious or controversial claims so long as they seem to reflect competent work and measured conclusions. The journals as a group are rather inclusive, yet they do impose substantial constraints upon what is acceptable discourse, far more exacting than what will be applied to informal discussion.[122] The retrospective literature, ranging from periodic review articles to textbooks and reference works, must be far more guarded. It aims to reflect the considered judgment of the disciplinary community as a whole rather than the claims of individual scientists. If a unified, consistent body of scientific knowledge, distilled through an extensive process of criticism, is to be found anywhere, this is where it should be. Yet such publications inevitably reflect the idiosyncrasies of their

120. Macomber 1968: 201.

121. Perhaps the most famous example of this is the divergence between the reference book version of the tautomeric form of guanine and what "everybody" knew to be the case, reported by Watson (1968: 120–22).

122. Ravetz 1970: 182–84.

authors, and in any case they are already behind the field even when they first appear. What "everybody knows" has already outdistanced them. Perhaps more important, they contain simplified, standardized versions of earlier work, overlooking or bypassing its obscurities and confusions and smoothing over its more esoteric subtleties. One finds here not a final, rational assessment of the credibility of recent results, but an attempt to adapt and prepare them for a different use in a different practical context. Such statements are gradually ceasing to be of concern in their own right and are accepted, rejected, or ignored for their possible contribution to a new field of controversy that has replaced them at the forefront of research.

We find, then, that there are no generally applicable standards of rational acceptability in science. There is only a roughly shared understanding of what can be assumed, what can (or must) be argued for, and what is unacceptable for any given purpose and context. Both purposes and contexts are quite varied and undergo significant transformations over time. They reflect the judgments of a community concerning what is credible and reliable in the context of their ongoing work. That shared project and the concerns and interests it reflects, however marked by significant dissensus, provide the reference points for all criticism and evaluation in science. Scientists aim to contribute to the project marked out by the actual work of their colleagues. It is therefore a shared reference to "what one does" in this shared field of practices that provides the basis for this variety of assessments and standards. The point is not that scientific communities determine what is true or false. Instead, two points are being made. The first is that the concerns, practices, and utterances of such communities provide the background against which statements emerge as intelligible and acquire epistemological status. They determine what can be true-or-false rather than what is true and what is false. This point is clearly formulated in Heidegger's discussion of truth in *Being and Time* and in Wittgenstein's references to the semantic role of forms of life, but it has still not been fully understood and assimilated in contemporary philosophical discussion.[123] The second point is that the epistemological status of scientific statements is not simply bivalent. Scientists employ the terms "true" and "false," to be sure, but they represent different standards of evaluation and serve different purposes in different contexts. In any case, scientists employ a host of epistemological modalities ("the data suggest that p," "X has shown that p," "it has been reported that p," "p," or just assuming that p without needing to

123. Heidegger 1957: 212–30, 1962: 256–73; Wittgenstein 1953, pars. 241–42.

mention it) whose many variations reflect the shifting status of a statement as it moves through various scientific situations and fills a variety of functions. But truth and falsity *simpliciter* do not seem to play a significant role in scientists' evaluations.

Conclusion

The argument of this chapter has been rather long and involved. I began with the aim of providing an alternative to the "theory dominance" of post-Quinean theoretical hermeneutics. The strategy was to criticize Heidegger's early proposal that the sciences in part be exempt from his hermeneutics of practice. The first result of this criticism was to show that theory-dominant views overlook the engaged, opportunistic features of scientific research. Practical considerations govern the development and evaluation of scientific work throughout, since scientists are more concerned to advance their work than to accumulate results independent of the social context of research activity. Even scientific theories turn out to be loosely joined groups of overlapping, not always consistent extendable models rather than a systematic web of belief. Heidegger's claim that science thematizes its objects rather than using them does not avail here, because we have seen that science also has its equipment that withdraws from attention. What is thematic is the work at hand, but this work will eventually be transformed into equipment for further investigation if it survives in science at all.

Consideration of experimental practice reinforces this view. Experiment turns out to be more than just an appendage of theory. Experimental considerations and capabilities independent of theory play an important role in the development of science as theory and experiment influence one another dialectically. The local laboratory site turns out to be the place where the empirical character of science is constructed through the experimenter's local, practical know-how. The resulting knowledge is extended outside the laboratory not by generalization to universal laws instantiable elsewhere, but by the adaptation of locally situated practices to new local contexts. The laboratory can then be regarded as the site of a Foucaultian microphysics of power that eventually transforms the social world around it. What Heidegger once took to be the theoretical discovery of decontextualized objects was instead shown to be the standardization of highly specific theories, facts, instruments, and procedures into more general-purpose equipment that enabled such extensions of laboratory knowledge. Scientific discourse and its evaluation were then shown to belong to particular social contexts as well. Claims acquired their significance and epistemological status through negotiations within behaviorally self-adjudicating scientific communities. The standards of rational acceptability were shown to vary with the contexts within which such claims occur and the practical interests governing them. Scientific knowledge, then, must be understood in its use. This use involves a local, existential knowledge located in a circumspective grasp of the configuration of institutions, social roles, equipment, and practices that makes science an intelligible activity in our world.

Chapter 5

Against Realism and Anti-Realism

Discussions in the previous chapter of the construction of scientific phenomena and the contextuality of justification point us inexorably toward recent controversies over scientific realism and its apparent alternatives. This chapter will situate my account of science with respect to these controversies. This effort in turn will inform my interpretation in chapter 6 of the various attempts to demonstrate an epistemological or ontological difference between the natural sciences and the study of human beings.

Realism is typically construed as the view that scientific theories are true or false depending on whether the objects they describe (including unobservable objects like electrons and quarks) actually exist and have the characteristics the theories ascribe to them. Realists presuppose a correspondence theory of truth and insist that the way the (natural) world is does not depend on what we say about it in our theories. Otherwise their view could be dismissed as true, but trivial, if our theories were somehow responsible for their own truth. Realism thus entails the possibility that all our theories might be false; the world *might* contain rather different sorts of entities than we know of, behaving in unexpected ways. But most realists today are not skeptics with regard to science. They believe science actually does describe the world rather accurately. Their insistence that it might have been otherwise serves primarily to underline the significance of the success they attribute to the sciences.

These views have been opposed by two distinct kinds of anti-realist position. Empiricists (or instrumentalists) challenge the connection between truth and empirical adequacy. They claim that science is not trying to describe the way the world really is but only attempting to account for the available empirical evidence. Empirical adequacy is never sufficient justification for the truth of scientific claims, but empirical adequacy is all we actually need in science. (Alternatively, some empiricists are strict verificationists about truth; whatever cannot be empirically *shown* to be true or false cannot *be* true or false.) Empiricists usually distinguish sharply between those scientific claims that refer only to observable entities and those that refer to entities we could never observe. The former claims can be shown to be true or false; the latter cannot. This limitation does not mean such theoretical discourse is meaningless, because it might be reinterpreted in terms referring only to observables. But it does mean at the least that theoretical claims should be given credence only for what they tell us about things we can verify empirically.

Constructivists are anti-realist in a different way. They deny any such sharp distinctions between observational and theoretical statements. All our interpretations of the world are "theory laden." Their objection to realism is that the way the world is cannot be made intelligible without employing the vocabulary of scientific theories and the practices by which we attach it to the world. Unlike empiricists, who say that a correspondence theory of truth cannot be meaningfully applied to *some* of our scientific claims, constructivists argue that correspondence theories are altogether unintelligible. Such accounts of truth imply a relation between two things—scientific theory and the world uninterpreted by theory—but we can never encounter the latter or even make sense of it. In assessing the acceptability of a scientific theory, we always compare theories with other theories, never with an appearance of the world unmediated by theory.

At first glance it seems as if the treatment of science in the preceding chapters embodies a constructivist anti-realism. There were many important themes in my account that do not seem to accord with standard realist interpretations of science: the emphasis upon practice as the contextual background against which things manifest themselves; the willingness to accept that what there is changes as scientific practice changes; the denial of any disengaged, "theoretical" perspective from which we can neutrally assess our epistemic claims; and above all the treatment of theories as models we use to manipulate and control phenomena rather than just to describe and explain them.

There are, however, discordant themes that suggest this does not tell the whole story. Observation, which is so fundamental to empiricism, does not figure significantly in my account. I emphasized science as a way of acting on the world, rather than a way of observing and describing it. My challenge to theoretical hermeneutics seems to undermine constructivism as well. If research practices have a life of their own independent of theory and reveal the world in their own ways, then we must reject the claim that we have no access to the world except as theoretically interpreted. The emphasis upon experiment and research practice does not suggest the apparently idealist implications of constructivism. In emphasizing the projective character of science, and its *aim* to disclose the way the world is, my views even suggest some parallels to the carefully circumscribed realism advocated by McMullin.[1]

My position thus does not mesh well either with realism or with the varieties of anti-realism one commonly finds in the literature, and it may begin to look as if my account of science is simply confused with respect to realism, borrowing elements from incompatible positions. I will argue instead that there is serious confusion in opposing realism to empiricism and constructivism. The view of science I am proposing can sidestep the central claims of both realists and anti-realists without having to challenge our commonsense acceptance of most of the claims made within contemporary science. This challenge to realism as an issue may seem more initially plausible when one realizes the extent to which the antagonists over realism seem to differ among themselves about just what realism involves and why it matters. They almost all seem to think that their opponents or critics have somehow gotten lost, misconstruing their positions and failing to see just what is at issue in the dispute.[2] Indeed, there are already parallels to my criticism of the issue itself in the discussions of Rorty, Horwich, Fine, and Williams, and implicitly in that of Hacking.[3] My concern in this chapter is less to sort out these debates than it is to use them to clarify what is at stake in my account of science. In practice, this will mean showing how my views escape the criticisms raised against one another by realists, empiricists, and constructivists, since all three positions tend to be defended by demonstrating the untenability of the alternatives.

1. McMullin 1984.

2. It is instructive in this respect to read alternately the accounts of Putnam (1978, lecture 2 and part 3), Rorty (1979, chap. 6), Putnam (1982), Boyd (1984), Horwich (1982), Glymour (1982), Fine (1986), McMullin (1984), and Hacking (1983, chap. 1).

3. Rorty 1979, chap. 6; Horwich 1982; Fine 1986; Williams 1984; Hacking 1983, chaps. 1, 7, 8.

There is little doubt that in recent years realism has come to be the predominant ontological position among contemporary philosophers of science. It will be useful to begin by saying something about why this is so and what form the *arguments* for realism have taken. Perhaps the most powerful motives for the new interest in realism come from objections to various forms of empiricism and instrumentalism. It is widely recognized that attempts to define theoretical terms using only terms referring to observable entities and properties have failed. More important, the reduction of theoretical terms to terms referring only to observable entities has come to seem undesirable. The role of theory in guiding further research is given a new philosophical significance in post-empiricist philosophy of science, and theory has come to be seen as much more than the notational convenience the most radical positivists hoped to reduce it to. Theory tells us more about the world than could ever be known from observation alone.

The attack on instrumentalism goes deeper than just a renewed appreciation of the indispensability of theory, however. The very distinctions that underlie instrumentalist interpretations of science have come under attack. The attempt to draw sharp distinctions between observable and unobservable entities, properties, and processes has not fared well in recent philosophical discussion. On the one hand, our understanding and interpretation of what seemed to be straightforwardly observable entities has been alleged to be heavily laden with theory. As a result, the epistemological desirability of the distinction has also been challenged, since the seemingly brute, uninterpreted character of observational data helped motivate the attempt to reduce the more problematic concepts contributed by theory. On the other hand, the distinction has been attacked from the other direction: the line between the observable and the unobservable has been shown to be only contingent and pragmatic. Its contingency (we might well have had different observational capabilities) makes it hard to justify using it to decide what can and cannot count as real. Its dependence upon our technological capabilities, which of course have changed, makes it still more suspect. Do we see "through" telescopes? Microscopes? Why shouldn't we take at face value scientists' untroubled assertions that we "directly observe" the solar interior using the elaborate apparatus for detecting solar neutrinos we discussed earlier?[4] Realists tend to be sympathetic to the claim that there is a continuum from what is obviously directly observable to definitely unobservable theoretical posits, rather than a sharp line between them.

4. Shapere 1982.

Other continuities between ontologically suspect theoretical entities and those entities about which we harbor few doubts have further reinforced the tendencies toward realism. We use scientific equipment in research in ways not dissimilar to our ordinary use of tools, and this seems to suggest similar ontological commitments. Laboratory practice does not seem extraordinary or dubious, yet it relies upon our ability to manipulate and control supposedly "unobservable" entities as equipment. Ian Hacking graphically indicates the conclusion he drew from scientists' matter-of-fact employment of streams of positrons as tools for investigating other phenomena: "If you can spray them, they're real."[5] Others are more impressed by the continuity between the forms of *argument* we use to infer the existence of both observable and unobservable entities. Ernan McMullin notes that the same kinds of retroductive arguments are used to infer the existence of both "observable" but unobserved geological structures like tectonic plates or structural discontinuities within the earth and the unobservable particles postulated by theoretical physics. Why, he asks, should we give more credence to the one ontological claim than to the other?[6] Some anti-realists object to those supposed entities whose claim to existence relies almost entirely upon the success of a particular theory. Such entities are indeed vulnerable to changes in theory. But the realist has an obvious response here: We may be less entitled to give *credence* to theoretical claims of this sort, but why should we *interpret* them any differently? Such theories ought to be understood as claiming that certain kinds of entities exist, whether or not the theory is ultimately accepted. The force of these considerations has led even a revisionist empiricist like Bas van Fraasen to agree that theories must be *interpreted* realistically, though he insists that we should *accept* only those claims that are necessary to account for a theory's empirical adequacy. "The fact that we let our language be guided by a given picture, at some point, does not show how much we believe about that picture. When we speak of the sun coming up in the morning and setting at night, we are guided by a picture now explicitly disavowed."[7]

These criticisms of instrumentalism have pushed many philosophers of science toward what appears to be the central thesis of any view that calls itself realist. Realists want to detach their ontological views from our observational capacities and our confirmation prac-

5. Hacking 1983: 23.
6. McMullin 1984.
7. Van Fraasen 1980: 14.

tices. So they are committed to saying that the way the world (really) is may be entirely inaccessible to our scientific (or other) procedures of disclosure and confirmation. We might be entirely ignorant of the real structure of the world. (Of course, an empiricist like van Fraasen can accept this sort of view with equanimity; he would say so much the worse for the real, because it is not the business of science to try to tell us how the world really is apart from how we can observe it to be.) This is why Hilary Putnam's argument that the supposition that we are really "brains in a vat" is self-refuting[8] is intended as a criticism of realism. But as I noted earlier, the more intriguing recent defenses of realism, associated with the work of Richard Boyd, W. H. Newton-Smith, Ilkka Niiniluoto, an earlier Hilary Putnam, and others,[9] defend realism in order to promote just the opposite epistemic view, insisting that, at least in the "mature" sciences, our theories *are* converging on an accurate description of the way the world (really) is. This convergence is remarkable, because (they claim as realists) it might have been entirely different. Not only have these views been a focal point of recent discussions of realism, they also turn out to be useful in clarifying what is at stake metaphysically in my account of science, and they will figure centrally in what follows.

A composite version of this sort of "convergent realism" can be summarized in five central claims:[10]

1. The theoretical (and observational, insofar as these can be distinguished) terms of "mature" sciences generally refer successfully to entities in the world, where "reference" is usually construed in terms of some kind of causal theory.
2. The best theories in these sciences are approximately true.
3. The historical progress of these sciences largely consists of increasing approximations to the truth, building upon and extending the successes of past theories.
4. The reality these theories successfully describe is largely independent of our thoughts or theoretical commitments.
5. The preceding four claims are justified by an abductive argument to the best explanation, which explains the pragmatic success of the mature sciences. Realism, that is, can best explain why it is that

8. Putnam 1981, chap. 1.

9. Boyd 1973, 1980; Newton-Smith 1981; Niiniluoto 1977, 1980; Putnam 1978, lecture 2 and part 3.

10. Similar summaries of convergent realism can be found in Boyd (1984) and Laudan (1981).

theories in these sciences are such reliable instruments for prediction, and perhaps only realism can provide an adequate explanation for such success.

Such an account constitutes a powerful reaction to the initial constructivist developments in post-empiricist philosophy of science. It reinstitutes claims of cumulative progress in science and ties such progress to a correspondence theory of truth. It rejects the ontological relativism prominently associated with post-empiricism, and in doing so it replaces the Fregean theory of reference that both empiricists and their constructivist critics had presupposed. Perhaps most striking, it places philosophical arguments for realism on a par with scientific claims themselves. One argues for realism the same way one might argue for the existence of quarks, as the best available explanation for the observed phenomena within a particular domain. Truth and reference become explanatory terms that make a difference to what we can reasonably say about the natural world. By the same token, anti-realism becomes *scientifically* unacceptable if this argument holds; the commonsense acceptance of scientific claims as guides to everyday practice is supposed to require such a robust metaphysical realism for explanatory support.

Convergent realism as represented in theses 1–5 has been subjected to a number of powerful criticisms. Larry Laudan has challenged the connections alleged to hold between the realists' notions of reference and approximate truth and the pragmatic success of science and has questioned the historical tenability of realist interpretations of the relations between successive theories in a given domain.[11] Arthur Fine, Richard Rorty, Bas van Fraasen, and I, among others, have argued that the abductive arguments for realism do not provide good reasons to accept it.[12] Fine has also claimed that an insistence on the explanatory indispensability of realism conflicts with the central role of anti-realism within the recent history of physics: as a scientific hypothesis in physics, realism has arguably been retrogressive.[13] It is

11. Laudan 1981.

12. Fine 1986; Rorty 1979, chap. 6; van Fraasen 1980; Rouse 1981.

13. I am inclined to accept McMullin's response to *this* argument of Fine's. McMullin argues that realism requires us only to accept the structural ontology of the best available theories in science. If that structure involves nonclassical indeterminacies, then the realist will accept that this is the way things are. In this sense Bohr's work is no more anti-realist than Einstein's. There is, of course, a sense in which Bohr's interpretation of physics is anti-realist, but this is not the sense that is usually at issue in *general* discussions of scientific realism (as opposed to interpretations of particular scientific fields, such as quantum mechanics).

not my aim here to review these arguments, although I am generally inclined to accept them as objections to convergent realism. Rather, I propose to use this robust version of realism as a foil for interpreting the view of science I have been developing. As Fine has suggested, realism acquires its hold on us much less from the positive arguments adduced for it than from the undesirability of the apparent consequences of denying it.[14] The view I am proposing certainly denies or undermines the significance of many of the central theses of convergent realism. The question that must be asked then, is whether the consequences of that denial are acceptable. The issue is not whether and how I might object to convergent realism, but how a realist might object to my account. Are there adequate responses to the objections a realist might raise to practical hermeneutics and the understanding of science as primarily local, existential knowledge?

I will treat this question in four parts. The next section of the chapter will concern the argument that realism is necessary to *explain* the pragmatic success of the natural sciences. In the second section, I will show that the criticism of instrumentalism leaves my views unscathed; indeed, I will give my own reasons for rejecting instrumentalist interpretations of science. I return to realism in the next section and show that the first three claims of convergent realism concerning truth, reference, and progress are perfectly consistent with my account once "truth" is interpreted in a deflationary way. My deflationary acceptance of these claims does, however, require that I challenge the realist's insistence that the way the world is, is independent of our theories and practices. In the final section I give a parallel deflationary account of "existence" and "reality" that enables me to deny the realist's claim without having to endorse the more counterintuitive claims of constructivism, which suggest that we are somehow responsible for the way the world is. The result, I believe, is a view that undermines the presuppositions that make it appear something is at stake in the opposition of realism and anti-realism.

The Failure of Convergent Realism

The most striking direct argument for realism is the claim that only a realist can explain the general success of the sciences. As Putnam once put it, only realism can avoid regarding the success of science as an inexplicable miracle.[15] What is the success supposedly in need of

14. Fine 1986.
15. Putnam 1978: 19.

explanation? As Larry Laudan has pointed out, it cannot be the satisfaction of stringent and well-established conditions of confirmation, for no such conditions have been found that both avoid obvious objections and can be satisfied by actual scientific theories.

> One would like to be more specific about what success amounts to, but the lack of a coherent theory of confirmation makes further specificity very difficult. Moreover, the realist must be wary—at least for these purposes—of adopting too strict a notion of success. . . . What he wants to explain, after all, is why science in general has worked so well. If he were to adopt a very demanding characterization of success (such as those advocated by inductive logicians or Popperians), then it would probably turn out that science has been largely 'unsuccessful' (because it does not have high confirmation), and the realist's avowed explanandum would thus be a nonproblem.[16]

Also, if there were such a direct argument for the truth of scientific theories, the abductive argument favored by convergent realists would be unnecessary. Nor can the success of science be understood as the general reliability of a methodology of abduction, of extending well-confirmed results into new domains while preserving their previous success. The problem is that such a methodology is *not* generally successful. More often than not, attempts to build upon previous results by extension fail. Science is just more difficult than that, and any argument aiming to show that such tinkering with well-confirmed theories will generally lead to successful extension has aimed too high to be true to scientific practice. Thus, Arthur Fine concludes: "The idea that by extending what is approximately true one is likely to bring new approximate truth is a chimera. . . . The problem for the realist is how to explain the *occasional success* of a strategy that *usually fails*."[17] One cannot even say that the predictions licensed by currently accepted theory generally succeed, unless one allows for the many failures that intervene between an initial prediction and the final achievement of a reliable experimental verification of it. This limitation was the point of my argument in the preceding chapter; successful prediction in science is as much a matter of practical manipulative skill in the laboratory as of prior possession of correct theory. This is particularly true when one realizes the extent to which the theory itself undergoes subtle change through the attempt to attach it to the world through experiment.

16. Laudan 1981: 23.
17. Fine 1986: 119.

It is therefore generally agreed that convergent realists have in mind a more rough-and-ready, pragmatic notion of success. They are trying to explain the fact that bridges do not fall down, the lights do go on when you flip the switch, and experiments do work (eventually). What this rough-and-ready notion sacrifices in precision it gains in plausibility, since very few philosophers (or anyone else) are likely to challenge the success of science in some such sense. But this gain does not help the realist argument, both because there are ways to account for this kind of success without appealing to a convergent realism and because the appeal to realism itself does not provide the sort of explanatory underpinning its advocates are calling for. We can begin with the latter point.

The failure of convergent realism as an explanation for the pragmatic success of science has been widely noted.[18] The basic problem is that the realists want to use the same abductive appeal to unobservables (in this case, noumena rather than quarks or curved space-time) that is characteristic of many scientific theories in order to underwrite the success of those very theories. The argument against this procedure can be put in the form of a dilemma. Either the pragmatic success of our best scientific theories stands on its own without need of any further explanation or else the purported explanans, the real existence of the entities referred to in the theories, approximately as the theories describe them to be, is equally in need of explanation. The realist cannot explain, that is, why the abductive appeal to "real" entities causally related to our use of certain theoretical terms succeeds in underwriting the success of our best theories. Previously, of course, we explained the pragmatic successes of our theories by appealing to the theories themselves. The added efficiency of high-pressure steam engines is explained by the second law of thermodynamics; the covariation of certain phenotypic traits in *Drosophila melanogaster* is explained by the locations of the relevant DNA sequences on the chromosome; that the lights go on when I flip the switch is explained by some electromagnetic theory and the fact that the switch closes the circuit. It is not clear what realism is supposed to add to such explanations other than a certain percussive emphasis;[19] for if we take the realists' empirical pretensions seriously, they are doing exactly the same thing on a metatheoretical level. Arthur Fine has succinctly captured the methodological principle that shows why the realist ex-

18. See Rorty 1979, chap. 6; Laudan 1981; Rouse 1981; Fine 1984b.

19. Fine (1986: 31, n. 20, crediting John King) uses this term to refer to the table-thumping emphasis that realists sometimes place on the word "really."

planans cannot do the work demanded of it: "Metatheoretic arguments must satisfy more stringent requirements than those placed on the arguments used by the theory in question."[20] That is exactly what these arguments fail to do, for the realists even boast that their arguments are methodologically on a par with those they seek to justify.

The failure of realism as an explanation of the pragmatic success of science is not troubling, however, unless one can make a case that this success is in *need* of explanation. Realists have not been entirely forthcoming in accounting for why such explanation is called for. The most illuminating comment, often cited approvingly by others, is Putnam's claim that non-realist accounts of science make its pragmatic success a miracle.

> If there are such things [as electrons, curved space-time or DNA molecules], then a natural explanation of the success of these theories is that they are *partially true accounts* of how they behave. . . . But if these objects don't really exist at all, then it is a *miracle* that a theory which speaks of gravitational action at a distance successfully predicts phenomena; it is a *miracle* that a theory which speaks of curved space-time successfully predicts phenomena.[21]

This claim might be plausible if we developed and adopted scientific theories on the basis of epistemic considerations that had little to do with pragmatic success. But clearly this is not the case. We saw in the preceding chapter that scientific theories are developed to model particular empirical situations we know how to deal with and that, conversely, our experimental manipulations are developed with the aim of increasing their reliability. The dialectic of experiment and theory is governed in large part by the feedback of pragmatic success and failure.

If further explanation of the pragmatic success of science was called for, then, the most appropriate strategy would be an argument by analogy to natural selection. It is important to remember that most new formulations or extensions of theory, and most new strategies for experimental manipulation, do not work, or at least do not work in their initial versions. What primarily governs the resulting modifications is pragmatic reliability. This consideration is especially true of the standardized versions of theories, calculations, and experimental procedures and materials, which are preserved within the developing corpus of accepted scientific knowledge. Those theories and pro-

20. Fine 1986: 114.
21. Putnam 1978: 19.

cedures that do not exhibit such reliability are generally modified or discarded. The point is not that pragmatic reliability is the only criterion governing the acceptance of theories or even of laboratory procedures. Such considerations as convenience, elegance, and explanatory power are certainly important. But pragmatic reliability is a very important desideratum, and while there is a substantial philosophical literature that quite rightly insists theories are rarely abandoned because of predictive failures alone, such failures are certainly a focal point for attempts to improve theory. All things being equal, empirically successful versions of theories tend to be preferred to less successful ones. Thus it should not be surprising that the overall development of scientific practice through time exhibits empirical success. That is what we should expect from a practice that systematically aims to achieve and preserve such success. One no more needs to refer to the "real" reference and truth of theoretical terms and sentences to explain this effect than one needs to appeal to approximation toward ideal species to explain the successful adaptation of organisms to their environment. "Success" simply reflects the criteria by which unsuccessful theories (or populations) are eliminated.

The realist might still raise two objections to such a response, however. On the one hand, an argument modeled upon natural selection in this way does not tell us why some theories are successful and others are not; it only explains why scientists generally succeed at telling the difference. This objection, I believe, is the sort of concern Putnam had in mind when he stressed the role of realism in explaining the reliability of our learning procedures.[22] Why did the Bohr model of the hydrogen atom fail and the later quantum mechanical models succeed? Because, the realist wants to reply, the hydrogen atom really is a proton/electron quantum mechanical system and not two determinate particles, one orbiting around the other. But in making this move, the realist ignores one of the fundamental methodological moves of the scientific revolution, which is to rule out the question of why the universe has just the structure it does. Samuel Todes puts this point elegantly: "Theory, of course, makes the internal relations in the body of evidence taken as a whole remarkably lucid. . . . But a price is paid for this advantage, a price which goes generally unnoticed. *What* the facts are is made luminous by theory. But *that* these are the facts is plunged by theory into a darkness just as extraordinary as the light shed on their nature."[23] This concern has

22. Putnam 1978, part 3.
23. Todes 1969: 16.

more commonly been applied to teleological arguments and cosmological arguments for the existence of God. But it applies with equal force to the realist's abductive arguments. The realist is trying to provide a scientifically respectable answer to a question that science has for good reason ruled illegitimate. Once we have given the best available scientific account of a phenomenon, there is nothing more to be said about why that account is successful (except, of course, to situate it within a more comprehensive theory that encompasses still more phenomena; but the realist argument does not claim to do that).

The other objection to this response to the realist abduction is that the selection argument does not explain the differential success of various sciences. I will have much more to say about alleged differences between the natural and human sciences in the next chapter. But the argument here is much broader in scope. The claim is that the selection argument cannot explain why successful variants arise at particular times and at differential rates in different scientific fields. Realism can presumably explain these differences by referring to successful abduction of the real structure underlying the phenomena to be explained. But apart from the difficulty discussed above of whether the realist's argument strategy has any explanatory force at all, the realist explanation is not very helpful here. There is presumably a crucial turning point in each field (according to the realist) when one first latches onto a vocabulary that successfully refers and acquires an approximately true theory within the domain of that vocabulary. Thus the realist may well explain the empirical progress exhibited in moving from phlogiston chemistry to the oxygen theory of combustion by saying that for the first time chemists had a vocabulary that matched up reasonably well with the entities really involved in chemical reactions. But once this point is reached, it is not clear how one explains further differentiation. The realist argument says nothing about how one goes about improving one's approximations to the truth and why some fields make more progress than others in doing so. From the realist point of view, all mature sciences (all sciences that have a referring vocabulary and approximately true theories) are on a par in this respect. The realist explanation of success could not make further differentiations unless one had available a means for discovering degrees of approximation to the truth. But it is hard to see how this assessment could be done without some kind of direct access to the truths against which the various approximations are to be measured. And such direct access would at one fell swoop make superfluous both the approximations and the realist's abductive strategy.

Certain non-realist strategies (e.g., practical hermeneutics and the new empiricism) can by contrast avail themselves of some powerful explanations for some of the differentia of scientific success in various fields. By appealing to the development of practical capabilities that open new fields of phenomena to investigation, or that disclose previously known phenomena with greater precision, one can understand much of the success and the occasional stagnation of various fields of research. Consider genetics in the twentieth century. There was one burst of discoveries associated with mapping the chromosomes of *Drosophila melanogaster,* which involved new cytological techniques, the detailed study of phenotypic/genotypic correlations in a single organism, and the use of X-ray induced mutations as an investigative tool.[24] The study of bacteria and bacteriophages introduced new possibilities for investigation, and with them a burst of productivity in the field. Watson and Crick's discovery of the structure of DNA revitalized genetics in a related way: it eventually opened the biochemical structure of the gene to investigation.[25] Recombinant DNA techniques did for molecular genetics what X-ray mutations had done for chromosomal genetics, giving scientists access to new phenomena, new ways to manipulate and control the things they made manifest, and therefore new possibilities for articulating theory in relation to the world. Similar points could be made about the role of higher-energy particle accelerators in stimulating progress in particle physics. Science is often led by technical developments that open new opportunities for experimentation and theorizing rather than just by new (approximately true) theoretical insights themselves. Realists also have recourse to these kinds of explanation, of course. The point is not that realism cannot recognize the role of technical and practical developments in stimulating scientific progress, but that once their importance is recognized, one feels less of a need to appeal to realism to understand why such progress occurs. Often, of course, issues about differential progress in various fields of science are raised not as general philosophical questions but as a specific concern about the particulars of those fields. Here no general answer will be satisfactory, certainly not the simple realist response that progressive fields benefit from the right vocabulary and (approximately) true theories. For that answer does not help someone who is concerned about what to do in a field that is not progressing in a satisfactory way.

24. Allen 1975: 57–68.
25. Allen 1975: 205–28.

Why I Am Not an Instrumentalist

The attempt to explain the success of science has been the central motif in the most prominent recent defenses of realism, but it may not be the most compelling reason for recent philosophical interest in realism.[26] Many philosophers have regarded realism as the only defense against instrumentalism, verificationism, and the allegedly relativist and historicist tendencies of much post-empiricist philosophy of science. As a result, their defenses of realism have focused upon the criticism of the alleged alternatives, coupled with arguments to show that realism is not unintelligible or incoherent and that it can therefore save us from philosophical disaster.[27] The force of this argument comes from the fact that almost no one is willing to challenge the practical validity of the results of most scientific inquiry. By "practical validity" I mean two things: their reliability in the course of our ordinary behavior and the lack of good reasons to prefer any alternative version. We all feel compelled to take account of the entities and processes scientists tell us are there, and we almost never have a plausible substitute for what they say (or any desire to look for one). This compulsion is as true of an empiricist like van Fraasen, who insists that the empirical adequacy of a theory is good reason to use it (but not to believe in the real existence of any unobservable entities it postulates), as it is of the most committed realist. Fine summarizes this attitude this way: "If the scientists tell me that there really are molecules, and atoms, and psi/J particles and, who knows, maybe even quarks, then so be it. I trust them and, thus, must accept that there really are such things, with their attendant properties and relations."[28] Realists have tried to argue that if one is unwilling to accept realism, one must reject these seemingly unexceptionable claims. But to use this argument is to impose a false dilemma upon anti-realism, one that depends upon equating practical validity with a realist construal of truth as correspondence.

Various attempts have been made recently to distinguish between our ordinary acceptance of the results of science and the claims of scientific realists.[29] All of them have hinged in one way or another on the notion of a "deflationary" or redundancy account of truth. Deflationary accounts insist that essentially everything we need to know

26. Arthur Fine has made a similar claim (1986: 126–7).
27. For an example of such an approach, see Newton-Smith (1981).
28. Fine 1986: 127.
29. Horwich 1982; Fine 1986; Williams 1984.

about truth is expressed by the Tarski schema, " 'Snow is white' is true iff snow is white." This schema is a purely semantic conception of truth, which says nothing whatever about what it is for snow to be white, except perhaps what can be expressed in an equally pure semantics of reference, such that "In the sentence 'Snow is white,' 'snow' refers to snow, etc." Truth is to be regarded as the most basic semantic notion, and it is not to be confused with epistemological issues. This definition enables us to distinguish a "semantic realism" that unproblematically accepts the truth of most scientific claims (in the sense of a willingness to assert them) from the more robust realism that invokes a correspondence theory of truth to supplement the Tarski schema as an account of what it is for snow to be white. As Fine has insisted, such a semantic realism constitutes a kind of "core position" that is almost never at issue in debates over realism. The only question, he notes, is what (if anything) is to be added to this core position.[30]

The principal reason to distinguish this core position is to argue that one need not supplement it with either a strong realism (e.g., convergent realism as I have described it) or some form of instrumentalism or verificationism. This distinction severs the connection realists have often taken for granted between anti-instrumentalism and a robust realism. My account of science shares this much with such semantic realists: it rejects instrumentalism, it accepts a deflationary account of truth, and it denies any reason to postulate some stronger form of realism. We need to explore these points, to consider just where I take issue with convergent realism, and then to ask what my account does add to the core position of semantic realism.

Let us begin by considering the response of a practical hermeneutics of science to instrumentalism. Why is my discussion of science in chapter 4 inconsistent with traditional versions of linguistic empiricism and instrumentalism? To begin with, I accept the arguments that the distinction between the observable and the unobservable is a pragmatic one that has no ontological implications. I do so with equanimity, because observation and observability do not play an important role in my account of science. This lack of importance has to do with the most fundamental difference between my understanding and most other accounts of science. Philosophers of science have traditionally thought of science as a system of representation, whose aim is to describe accurately a world that is indifferent to how it is represented. Observation was important because it provided the only link between the world

30. Fine 1986; see also Williams 1984.

as we represented it to be and the world itself. Only in sense experience does the world impinge upon us in a way that constrains the possibilities for representing it. Thus we have Quine's claim as typical: "Whatever evidence there is for science is sensory evidence."[31] Things look considerably different from my perspective. The question is not how we get from a linguistic representation of the world to the world represented. We are already engaged with the world in practical activity, and the world simply *is* what we are involved with. The question of access to the world, to which the appeal to observation was a response, never arises.

The important categories for characterizing the ways the world becomes manifest to us are therefore not the observable and the unobservable. We must ask instead about what is available to be used, what we have to take account of in using it, and what we are aiming toward as a goal. It is not the physical constraints of our sense organs that determine these, but the behavioral adjudication of a community of practitioners. This position bears some obvious affinity to Hacking's experimental realism. He, after all, has said:

> The best kinds of evidence for the reality of a postulated or inferred entity is [*sic*] that we can begin to measure it or otherwise understand its causal powers. The best evidence, in turn, that we have this kind of understanding is that we can set out, from scratch, to build machines that will work fairly reliably, taking advantage of this or that causal nexus. Hence, engineering, not theorizing, is the best proof of scientific realism about entities.[32]

Hacking is careful to say that this is not the only acceptable evidence for existence claims, but he does not say much about the force of other sorts of evidence. The view I hold about this issue is holistic: one cannot manipulate one kind of thing for some purpose without understanding and using many other kinds of things. The things that are manifest within the practical context of scientific research form a complicated network, such that the manipulation of one thing requires us to take account of others, whose existence is thus implicated in those practices. The differences between what we actually manipulate, what we must take account of in doing so, what we represent in theories that inform or guide our manipulations, what we encounter as obstacles, and what our activities lead us to speculate about form a continuum. We will give different degrees of credence to the various

31. Quine 1975: 75.
32. Hacking 1983: 274.

existence claims that result from these different manifestations, but we do not interpret them differently as existence claims.

Van Fraasen's more sophisticated empiricism takes a different tack concerning the significance of observation with respect to realism and requires different arguments in response. Van Fraasen has no qualms about interpreting scientific claims realistically, even those that refer to unobservable entities. He also grants that what is observable does not circumscribe what can count as real; there are undoubtedly real things that can never be made manifest to our observation. He then accepts that the limits to observation are not fixed once and for all but are themselves determined by the appropriate scientific theories. He insists, however, that scientific claims are constrained only by that subset of the real that is observable, whatever its precise membership turns out to be. We assess scientific theories with respect to their empirical adequacy, not their truth, and as a result their success licenses us to claim only that they are empirically adequate.[33] Van Fraasen concludes that "even if observability has nothing to do with existence (is, indeed, too anthropocentric for that), it may still have much to do with the proper epistemic attitude to science."[34]

Van Fraasen's account thus avoids many of the standard objections to instrumentalism, usually by conceding the realist's point with respect to ontology while maintaining an empiricist view on epistemic grounds. But from my point of view there is a twofold problem with van Fraasen's position. The most obvious problem is that he maintains the traditional emphasis upon observability as a philosophically significant category. I am sympathetic to his emphasis upon empirical adequacy as the most fundamental epistemic criterion for assessing the success of science. But I regard the empirical adequacy of science as including the reliable ability to manipulate and control phenomena in the laboratory, whether or not the entities we manipulate are themselves observable. Our assessment that the equipment is working effectively, and that we understand what is going on in it, involves more than just an inference from the observable data that may be its output. Our practical grasp of the equipment and the theories that are integrally involved in its design and operation are quite relevant to any judgment about the empirical adequacy of the scientific practices in question. Indeed, our confidence that the observable data are significant, and not just artifacts, depends upon such practical understanding. Van Fraasen is concerned with the empirical adequacy of

33. Van Fraasen 1980.
34. Van Fraasen 1980: 19.

theories, meaning their successful prediction of observable data. I am concerned with the "empirical adequacy" of scientific practices, meaning their ability to manipulate and control phenomena reliably and ultimately to extend that control outside the laboratory into the world. Thus I find observation, in its philosophical sense of looking at and recording isolated data, to be less basic than practical circumspection.[35] The latter, to repeat, is responsible for any significance we might confer upon the former on any specific occasion.

My second objection to van Fraasen is that in an odd way he concedes too much to the realist. Van Fraasen abandons ontology to the realist but insists that science is concerned to give not a true account of what exists, but only an empirically adequate one. I will claim that science *is* concerned with what really exists, and that restricting its epistemic concern to what is observable does violence to what scientists do and what their aims are. But I will argue later in this chapter that this restriction is consistent with the claim that what there is does depend upon what can be made manifest in our practices. There undoubtedly are things we cannot observe in any straightforward sense. But what there is cannot be separated from what can become manifest in the context of what we do. There is an essential connection between being and manifestation. Empiricists have recognized this, but they went wrong in construing manifestation too narrowly in terms of sense experience. Van Fraasen went wrong in overreacting to the inadequacies of earlier empiricists' ontological scruples by giving up the idea that science is concerned to reveal what exists, as it exists.

There is also another powerful tendency toward realism in the views I am presenting here. Instrumentalism requires a distinction between the ordinary objects manifest in our everyday lives and the objects introduced by scientific theory. But an important part of my account in chapter 4 was the discussion of the ways objects first encountered in scientific research get introduced to the world of everyday life. I want to argue that the way the world is has (really!) changed as a result of science and that what were once "theoretical entities" in science became objects in the everyday world. This view is incompatible with one that gives a lesser ontological status to theoretical entities of science than to "ordinary" objects (unless one could somehow justify the rather implausible claim that fictional or merely instrumental objects become real when introduced into the everyday world).

This view, that science introduces new objects into the world, goes

35. Hacking 1983, chap. 10.

back at least to Husserl's *Crisis of European Sciences and Transcendental Phenomenology*, and it has been discussed in some detail by more recent authors as a "horizonal realism" or "phenomenological realism."[36] Heelan and Compton describe this view in terms of science making new sorts of objects perceptually manifest, and they therefore emphasize the importance of a perceptual hermeneutics for understanding science. Heelan notes: "In my account, *reality* is taken in a different way: it is exactly what worlds make manifest (or purport to make manifest) to human perceivers; consequently, science to be realistic must have as its primary goal the exhibiting of reality structures not accessible to prescientific perception."[37] My disagreement with Heelan on this point is perhaps no more than a difference in emphasis. I argue that things become "perceivable" in the course of our practical dealings with them and that the key effect of science is to enable us (but also compel us) to take account of them in our practices. Heelan emphasizes the perceptual payoff of such manipulations. I think there is value in stressing practice instead of perception here. Most fundamentally, this emphasis establishes a continuity between our understanding of how things become manifest within scientific practices and how those things affect the larger social context within which science is situated. In Latour's happy phrase, it dissolves the boundaries between the inside and the outside of science and focuses upon how scientific knowledge is extended beyond the laboratory walls within which it is developed.[38] The emphasis upon practice also avoids the confusions that result from traditional empiricist assumptions about perception. Heelan is careful to stress that perception is hermeneutic, that it is always situated within "a system of behaviors or praxis,"[39] and that it is embodied not just within our sense organs or even our whole bodies, but also within the equipment with which we engage the world.[40] But though this view certainly has antecedents (e.g., Merleau-Ponty), it is at odds with common philosophical assumptions about perception and perceivers. Placing such a central role upon perception insufficiently emphasizes the radical change that occurs when we understand science as a practical engagement with the world. Once this difference in focus and emphasis is noted, Heelan's and Compton's positions on the issue of realism have

36. Husserl 1954, 1970; Heelan 1983, part 2; Compton 1983, n.d.
37. Heelan 1983: 174.
38. Latour 1983: 153–56.
39. Heelan 1983: 176.
40. Heelan 1983; see especially chaps. 10–11.

much in common with mine. But I do think the differences are important.

The crucial point concerning instrumentalism, which Heelan, Compton, and I share, is that we reject any distinction between the theoretical entities we deal with in scientific research and the garden-variety objects we encounter in everyday activities. Science is philosophically on a par with ordinary activity, and the same is true of their respective objects. If one wishes to interpret me as an anti-realist about theoretical entities in science, one must regard me as comparably anti-realist about ordinary objects as well. But mine will not be an anti-realism offering an alternative ontology to that of ordinary objects and theoretical entities or an account of truth incompatible with a deflationary employment of the Tarski schema. Classical instrumentalism is just a realism with scruples: instrumentalists are realists about observable entities and not about others. Whether you want to call practical hermeneutics realist or anti-realist (a case could be made for either, depending upon how you construe the issue),[41] it treats observables and unobservables alike.

Deflating Truth, Reference, and Progress

I can now begin to develop more systematically the ontological views about science that I have alluded to several times in discussing instrumentalism and van Fraasen's constructive empiricism. I will do so by reflecting upon the first four claims I attributed to convergent realists, those concerning reference, truth, progress, and the person independence or theory independence of the real.

The first three issues—reference, truth, and progress—form a package that can be considered jointly. I have already indicated that I accept a deflationary account of truth. Pretty much all I want to say about the semantics of truth is contained within the Tarski schema of partial definitions. Truth plays no explanatory role, and the Tarski definitions do not need to be supplemented by a correspondence, coherence, or any other theory of truth.

But my reasons for accepting deflation are perhaps somewhat different from those of most deflationists. If I think that there are no substantive issues concerning truth, it is because I see the issues situated on the right side of the Tarski equivalence instead of the left.

41. For example, McMullin (1984) might call it realist, I think, while Boyd (1984) would clearly regard it as anti-realist.

There are no fundamental philosophical issues peculiar to the concept of truth, for they are the same issues that arise concerning how things have any determinations at all. What is it for snow to be white? Realists think that things have such determinations independent of our practices, desires, and beliefs. So the realist answer to my question is that snow is white iff snow really is white, regardless of what we think about or do with it. I think that what there is cannot be entirely separated from who we are and what we do. And this position suggests that the really fundamental issue between me and the realist concerns the fourth realist thesis, that reality is independent of our thoughts and theoretical commitments.

But before we turn to that issue in the last section, there are some considerations about truth that it may help to clarify first, for there are several significant reasons why even a deflationary account of truth might be thought not to enable us to say that our best-established scientific theories are true. Specifically, I want to consider some qualms expressed by Ernan McMullin and Nancy Cartwright[42] about taking scientific claims to be true, together with another apparent objection to scientific truth claims that appeared in chapter 4 above. Both McMullin and Cartwright are realists about the entities that mature, well-established sciences refer to in their theories (McMullin might prefer to say that he is a realist about the ontological *structures* postulated in such sciences). But they balk at saying that the fundamental theories of science are true, or approximately true. McMullin believes "truth" carries connotations of completeness and finality that are inappropriate to the enterprise of science. He has in mind not just the fallibilist recognition that our theories may well have to be changed to accommodate new data, answer new questions, and fit with further developments of theory in related domains. McMullin emphasizes the projective or horizonal character of science. Scientific theories are not put forward as final accounts of fully understood aspects of the world. They are essentially programmatic, indicating possible directions for exploration and serving as a resource for disclosing and assimilating new domains and new aspects of old ones. It is with this character in mind that McMullin stresses the metaphoric quality of theoretical language: "The language of theoretical explanation is of a quite special sort. It is open-ended and ever capable of further development. It is metaphoric in the sense in which the poetry of the symbolists is metaphoric, not because it uses explicit analogy or because it is imprecise, but because it has resources of suggestion that are the most immediate

42. McMullin 1984: 35–56; Cartwright 1983.

testimony of its ontological worth."[43] The fallibilist and exploratory quality of science is precisely what the convergent realist tries to leave room for in characterizing the truth of scientific theories as "approximate." McMullin worries that this characterization implies the wrong sort of claim. His realism does not commit us to saying we are already close to the final truth about some scientific issues. It says only that we have achieved a revealing insight into the real structures of the world, which promises still further disclosure if we pursue it.

Cartwright's concerns are somewhat different. She argues that explanation in science has a pragmatic character and that there is often a trade-off between explanatory power and precise, literal truth. We simplify our fundamental theories in order to encompass more phenomena within the scope of a limited number of models. This simplification sacrifices accuracy to obtain explanatory and calculative power. Nor is it just that we have to introduce approximations into our theories to save the phenomena. Our theoretical models may not even be consistent with one another. Different models of the same phenomenon may be useful for different purposes or in different contexts. There is a place for accurate phenomenological description in physics, since we want to know where and how far our explanatory models depart from the phenomena they deal with. But much of the business of science is to enable us to *do* more things with the phenomena: to calculate and manipulate them, exhibit interconnections among them, and so forth. For these purposes, the demand for phenomenological truth may be counterproductive.

But McMullin's and Cartwright's concerns tie in centrally with a practical hermeneutics of science. McMullin's emphasis upon the open-ended, projective character of science is exactly what I tried to capture in taking science to be first and foremost the practice of research. And I have already cited with approval Cartwright's account of theoretical models and the way they connect to the world. The question is whether such an account of science as a pragmatic, exploratory coping with the world is compatible with a deflationary semantic realism. I will argue that it is compatible; showing this compatibility will emphasize just how deflationary I think our "theory" of truth should be. As I have hinted above, my ontological views have some important parallels with empiricism. But I think that these parallels do not extend into the semantics of truth.

A deflationist account makes "true" a redundant predicate. Asserting "'p' is true" says no more than asserting p. But this does not mean

43. McMullin 1984: 36.

that "'p' is true" is equivalent to "I assert that 'p,'" even if the personal pronoun is replaced by a nonindexical expression. "I assert that p" makes a claim about me, which asserting p does not do (although it may provide the evidence for someone to assert that I assert that p). The semantics of truth do not, then, say anything about the material conditions under which the assertion of p is justified, or about the truth conditions of p (except in the limited sense that those conditions obtain whenever p holds). They say only that "'p' is true" and "p" are materially equivalent.

To understand when it would be the case that p is to understand something about the meaning of p. There is nothing mysterious about understanding the meaning of a sentence. This knowledge can be construed in public terms of appropriate use and its recognition. I understand a sentence when I know when and how to use it, when I can recognize its appropriate use by others, and when I can respond to it with appropriate behaviors (including the use of other sentences). The judgment of appropriateness derives from the self-adjudicating behavior of the community within which the assertion is to be understood. There is, of course, an interpretive circularity here, but not a vicious one: what allows a certain behavior (including the use of a sentence on a particular occasion) to count as appropriate is that it is (or could be) recognized as appropriate by competent members of the community; what counts as recognition is similarly socially adjudicated; and persons count as competent members of the community if they are recognized to understand in this way a sufficient portion of the community's sentences and behaviors. This view of meaning is similar to the one Horwich indicates as essential to semantic realism. He remarks:

> It is normally reasonable to attribute understanding—a grasp of the meanings of the sentences of a language (including unverifiable ones)—to anyone who displays the ability to use the language in accordance with community standards. . . . Moreover, we should describe [this ability] as *knowing how to use the expressions of the language, associating the right meanings with those expressions,* and in particular, *knowing what the sentences assert,* and therefore, given Tarski's schema, *knowing their truth conditions.*[44]

This theory of meaning is essentially the sort that Heidegger develops in division 1 of *Sein und Zeit.*[45] Horwich thinks that "understanding" as he characterizes it can ultimately be tied to some sort of internal

44. Horwich 1982: 184.
45. Heidegger 1957, 1962; see especially sec. 32.

psychological or neurological structure, whereas Heidegger takes it to be behaviorally construed throughout. But even Horwich notes that his internalist reading is quite inessential to it as a theory of meaning and therefore dispensable in construing the relation between meaning and truth.[46]

Such a theory of meaning enables me to preserve a deflationary account of truth while respecting McMullin's and Cartwright's scruples about claiming that scientific theories are true. This semantic realism is also consistent with my claims in chapter 4 that the criteria for the acceptability of a scientific statement vary considerably with the context in which the statement is made.[47] In this last case there is no difficulty with the fact that the same sentence will count as true in one context and false in another. This is not a case violating the law of noncontradiction, because the difference can be traced to a difference in the meaning of the two sentence tokens, despite their graphic similarity. This difference of meaning is nothing more than the fact that they count as true or false on the basis of different conditions obtaining in the world. Each is true or false according to the conditions that actually do obtain. But *which* conditions are to be related to the truth of each sentence token in a Tarski schema is determined by the behavioral responses of a community.

Similar points can be made about McMullin's and Cartwright's concerns. Cartwright worries that fundamental theories in physics offer a trade-off between explanatory power and exact phenomenological description. On the deflationary account of truth I am advocating, this trade-off is not to be understood as a conflict that places the *truth* of such theories in question. Rather, it shows that the conditions under which explanatory theories are understood to hold are less exacting than is the case with phenomenological accounts. This greater latitude of course does not mean that anything goes in these cases, since there are more or less definite constraints in each sort of case on the tolerance permissible in the match of theory and data.

McMullin's concern is that scientific claims are not understood to have the finality and completeness he and others associate with "truth." But the semantic interpretation of truth I am suggesting here is entirely neutral with respect to such finality. The conditions according to which a scientific statement is understood to be true are just the conditions under which it is appropriate to assert it. If, as McMullin quite reasonably claims, scientific statements are made in an open-

46. Horwich 1982: 200, n. 3.
47. Chapter 4 above, pp. 123–125.

ended way, suggesting a program for disclosing the world as much as a final statement about the way the world is, this openness will be reflected in the truth conditions that constitute their meaning. And their truth will depend upon whether the world is such as to be investigated fruitfully in terms of the program suggested in that statement. A Popperian might be concerned that this approach makes too many possible arrangements of the world consistent with the truth of a theoretical claim. But Popper's concerns are also an issue within the scientific community itself; a claim that turns out to be vacuously true will not normally guide research in any definite direction. So the somewhat open-ended conditions under which a programmatically interpreted scientific statement is judged to be true will normally rule out many prima facie possible states of the world, even if they cannot be specified in a rulelike way in advance. Darwin's claim that the diversity of species is due to a process of natural selection is a good example of this situation. His claim has undergone considerable amplification, qualification, and extension, but there are very few scientists today who would deny that *Darwin's* claim (and not just the more sophisticated selectionist positions that have since been developed) is true. It is this sense of truth that the deflationary account aims to capture.

I have been discussing a semantic deflation of the convergent realist's claims about truth, but parallel points could be made about reference and progress. What a term refers to is to be specified behaviorally by the self-adjudicating response of a community of practitioners who understand its use. It refers to whatever entity or kind of entity one is dealing with on occasions when it is appropriate to use the term. This view is very similar to "causal" theories of reference, when they are understood as a constraint upon interpretation rather than in a metaphysical realist way. That is, the causal chains used to specify reference are those we can recognize and attribute on the basis of our dealing with the world, not those that "really" occur regardless of our attributions. For the latter notion is unintelligible. Scientific progress according to my account does consist in building upon past successes and correcting past failures. But there is no criterion for assessing our ability to do so that does not refer to the aims pursuable within current scientific practice, including the equipment, practices, and roles that must be understood along with them.

Meaning and the Independence of the Real

We can now turn to the fourth thesis of convergent realism. Realists will undoubtedly argue that I have preserved our ordinary uses of "truth," "reference," and "progress" only by construing the real to be dependent upon our theories, practices, and standards. Where my view may seem to the realist to make the way the world is unacceptably dependent upon us is in my account of the connection between understanding and truth conditions (and similarly, the conditions for what counts as reference and progress). If to understand the meaning of a sentence is to understand the conditions under which it would be true, and if understanding is to be construed behaviorally in terms of knowing how to respond appropriately to it, then what can be true and what we can understand become internally connected. Only those conditions of the world that we can understand become connected to the truth of sentences. And this restriction seems to run headlong into the realist objection. For it seems evident to the realist that our sentences might be false precisely because of features of the world that are not accessible to us within the context of our current language and practices. This qualification is important, because it seems to explain both why the success of our interpretations is significant (it might have been otherwise) *and* how unsuccessful interpretation is possible. The realist believes that if there is any sort of internal interconnection between interpretations and what they interpret, as I am claiming, it is not clear how they could fail.

This concern to understand the possibility of failure partially accounts for the importance convergent realists have ascribed to explaining the pragmatic success of our practices. For what they are really trying to explain is the *difference* between those practices that succeed and those that do not. In saying that only realism can explain success, they are implicitly saying that any other view makes either success or failure unintelligible. Usually in their arguments they have emphasized the difficulty of explaining on instrumentalist grounds how science can be successful. If one accepts the (realistically construed) truth of observation statements but does not accept the realist's claim that our theories are (approximately) true and our theoretical terms referential, then the success of science at predicting true observation sentences may seem miraculous. But for realists the other side of the dilemma must be equally important. It must seem to them that when constructivists abandon a realist metaphysics altogether and claim that what is real is internally connected with our theories and practices, the consequence would be that those practices could never fail. For the realist, failure must be described in terms of our running up against a reality that is other than how we interpret it to be, and constructivism seems to make that impossible in principle.

The core of my response to the realist's objection is that the objection misinterprets the role of our practical dealings with the world in making meaningful both our interpretations and the things we interpret. I begin by noting that in saying that what is real is independent of what we do or say, the realist has a definite picture of the relations between our interpretations of the world and the world itself. Our interpretations say something definite about the world, which sometimes matches the way the world is and sometimes does not. This is the case not just with our sentences, but with our actions: sometimes what I pick up and (try to) use as a hammer is a hammer, and sometimes it is not. This difference is what accounts for whether the nail goes in or not. Any correspondence or fit between our interpretations and their intended objects is therefore contingent.

The problem with this picture is that it takes as already determined both the way the world is and our understanding of how our interpretations take it to be. The realist of course recognizes that we do not know in *advance* how the world is. But once we have some definite interpretations of the world, we can use them as the basis for our actions, which in turn test the adequacy of our interpretations. If our actions fail to achieve their aims, something must be wrong with the interpretations they were based on. If our actions succeed, this success of course does not entail that their underlying interpretations do accord with the reality they interpret. But if a wide variety of actions in differing circumstances generally succeed, the best explanation for their success is that those interpretations at least approximately accord with the way those objects really are. But where do we acquire our understanding of what our various interpretations do say about the world and of what would count as success in our actions? The realist needs to give some account of understanding such that we can understand how our interpretations take the world to be independent of how the world actually is. Otherwise the alleged independence of object and interpretation can never get off the ground. Sentences and practices do not have ready-made meanings, nor do they acquire meaning by convention. (How could the parties involved understand what they were agreeing to?) They acquire meaning only in their performance or use.

Performance and use are themselves intelligible only within the context of the various ends for the sake of which we perform actions and use equipment (including sentences). For the things we do to be meaningful to ourselves and intelligible to others, there must be something at stake, something that counts as success or failure in satisfying our ends. This prior understanding of what would count as success or failure is what Heidegger termed the preconception (*Vorgriff*) of every interpretation. To have such an understanding, one must understand many other things as well. Purposes and their satisfaction are understood together with the variety of equipment available for satisfying them, the ways that equipment is to be used, and so forth. As I argued in chapter 3, practical understanding is always holistic. One cannot understand any particular task or item of equipment (including a sentence) without some grasp of the whole configuration of things that makes up an intelligible world. "Meaning" is what it is that we understand; so the meaning of a thing is its place within such a configuration of purposes, practices, and equipment.

Our understanding of this configuration and the things that acquire meaning within it cannot be separated from the success or failure of our practices. My understanding of hammers is an understanding of what we use them for, what we use them with, and what ends are to be served in using them. My grasp of what counts as a hammer (the practical analogue to the truth conditions of a sentence) involves understanding what it is to hammer successfully and what sorts of things can be used successfully for hammering. We do not first abstractly specify the conditions for being a hammer and then look around to see what, if anything, satisfies those conditions. The conditions for being a hammer arise out of our actual dealings with hammers and out of the successes and failures of our attempts to hammer with them and thereby to satisfy the purposes for which we engaged in hammering in the first place. Hammers are those things that turn out to have been good for hammering, where hammering in turn is understood in terms of success or failure in satisfying certain purposes. This point can be made in a general way concerning both language and action. What a noun turns out to refer to is whatever is usually present (in the appropriate way) when the sentences in which we employ it contribute to the success of our behavior in achieving our goals. What counts as an x is whatever usually fits into our activities in such a way that by employing it in the right ways (behaviorally interpretable, though usually not fully specifiable) we can achieve the goals for which x's are appropriate means to employ.

The point in all this is that the realist's assumptions about the independence of our interpretations from what they interpret is violated in the very process of specifying the meaning of those interpretations. Realism takes the meaning of our interpretations and the way the world is as already fixed; the only question is in which cases the interpretations accord with the world. I am arguing that our dealings with the world come first. It is only through purposeful interaction with the world and the patterns of success and failure that emerge from it that our interpretations acquire meaning and the world becomes determinate. The real *is* what we manipulate, what resists us, what we notice, and what we take account of without ever taking explicit notice. To say that the things we deal with in our everyday actions exist is redundant in much the same way that saying the sentences I assert are true is redundant. Their existence is implied in my dealings with them, and it is redundant to add this claim to the interpretations already embodied within them.

But what about the realist's objections to any interdependence between our interpretations and what they interpret? The realist is concerned to show that the truth of our theories is not just an artifact of our research activities. If scientific research is to count as learning about the world, then it must not have such a strong influence upon what the "world" in the relevant sense consists of that we learn only about our constructive activities and not about the world as it was before that construction. As Horwich has pointed out, "there is a tolerably clear sense of 'independent' according to which some facts obviously are, and some obviously are not, independent of our existence and thought. 'If there were no human beings Disneyland would not exist, yet snow would nevertheless be white' is a view to which the proponents of almost any anti-realist position would subscribe."[48] We created Disneyland; we do not create snow (except at ski resorts), nor do we give it its color. If we could not sustain this distinction without adopting convergent realism, this failure would be a powerful argument in realism's favor.

Underlying the realist's objections to the various forms of constructivism is the belief that if, as a conceptual necessity, the world cannot be radically different from the way our theories construe it to be, then the apparent success of those theories is an artifact of their construction. And this consequence seems to a realist to be a reductio ad absurdem of any view that entails it. So the question we have to address is whether we can reject the realist's claim that the world is independent of the way we interpret it to be without entailing the unacceptable consequence that our theories and practices are only trivially successful because they cannot in principle fail.

My account of science in chapter 4 may seem particularly vulnerable to the realist's objection, since I argued that most of the phenomena scientists deal with are literally our own construction. They are in this respect more like Disneyland than like the whiteness of snow. But this claim does not really run afoul of the realist's objection.

48. Horwich 1982: 185.

For while I claim that the *phenomena* scientists investigate and attempt to manipulate and control are often created by us, this claim does not mean that the *entities* manifest in those phenomena are our creation. Electrical currents came into existence at the turn of the nineteenth century, and they did so because human beings constructed them. But this could be done because there were already electrons (even though there was little or no knowledge of electrons, depending upon how charitably you take discussions of the "electrical fluid"). The new scientists of electricity only made new arrangements of what was there, if you like. And of course nothing I have said challenges the claim that those laboratory arrangements (batteries, circuits, and electrical currents) were themselves real or that electrons turn out always to have had the capacity to flow in sustained currents under the right conditions. I have only said that these conditions themselves are a human artifact. Indeed, the construction of those conditions, like every other practical engagement with the world, is responsive to the capacities of the things we deal with. But these capacities emerge as intelligible only in relation to other things and the practices through which they are related.

It is the Cartesian image of privileged access to what we mean that fosters the illusion that our interpretations could be somehow "prior" to what they interpret. If the meanings of our interpretations were already transparently given, the claim that the world is always already interpreted might make some objectionable form of idealism an inevitable outcome. But we must look to the world to guide our interpretations of what we do and say. Instead of saying that we construct the way the world is, we could just as well say that the world shapes the meaning of our words and deeds. But it would be better to say that our interaction with the world takes precedence over any dichotomy between interpreting and the interpreted. This is what Heidegger meant by saying that we are "Being-in-the-world." Neither world nor our ways of being in it come "first." Each becomes determinate only in relation to the other.

Once this relationship is understood, we can see why the realist's concern is misguided. There are three points that must be understood together. The first point is the one that, by itself, makes the realist's objections perhaps look plausible. This is the claim that the linguistic and practical configuration within which a behaviorally self-adjudicating community finds itself determines what can count as a case of any predicate and thus opens whatever possible ways there are for things to be. Heidegger originally made this point with respect to equipment, where it is easier to understand. For there to be *hammers* presupposes that there is an intelligible activity of hammering that people know how and when to engage in. This knowledge requires a similar understanding of nails and boards and the purposes for which we fasten things together. Were there no such thing as nails (i.e., no *place* for them in our practices, as opposed to a de facto absence of them) there could be no hammers either. And of course the same is true of the beings who understand; without us, there could be no hammers either. In the former case, this assertion does not mean there could be no long, cylindrical wooden things with flat-ended metal pieces attached to one end; it means only that there could be no *hammers*. In the latter case, there could be nothing at all, not because we are in any way responsible for the being of things (e.g., causally), but because being is not separable from meaning, and meaning depends upon beings who can understand.

We can now see the importance of the argument in chapter 4. Heidegger thought that the objects of scientific theory were ontologically different from equipment in this respect but that the possibility of being something merely present-at-hand was ultimately dependent upon the being of equipment and the in-order-to-for-the-sake-of configuration within which it was intelligible. My argument suggested in contrast that supposedly present-at-hand things like electrons are ontologically no different from hammers. To be an electron is to belong likewise to a configuration of things we can intelligibly encounter in our purposive practices. What specifies a successful case of spraying a beam of electrons or cloning a DNA sequence is not fundamentally different from the specification of what counts as hammering with a hammer. Electrons have no simply specifiable function in the way hammers do, but they count as electrons in much the same way that hammers count as hammers. The characteristics that specify this have to do with the ways we (must) take account of them in what we do.[49]

We can now turn to the second of our three points. We have seen that the context of our understanding (our practices and the language we employ within them) does constrain what objects and properties can *count* as existent or nonexistent and what sentences can *count* as true or false; to this extent, the way the world is depends upon what we do and say. But within this context, which objects do exist, what

49. There are some obvious differences between typical items of equipment and typical objects of scientific research: the difference between being manufactured or found, and between being used as tools or encountered in other ways (e.g., as something worked *on* rather than worked with, as an obstacle to our work, etc.). But these differences are not relevant to the way *meaning* is constituted.

properties they have, and which sentences are true are entirely independent of what we do and say. To put it another way, the world is always interpreted, through language and practice. An uninterpreted world would be unintelligible. But the interpretation does not make the world the way it is; it allows it to show itself the way it is.

Let us return to our examples. Hammers are those things that reveal themselves to be suitable for hammering within a configuration of equipment, practices, and goals that holds a place for hammering as a significant thing to do. Without such a configuration, nothing could be a hammer, whatever physical characteristics it might have. But this constraint does not mean that we make things what they are (except in the ordinary sense in which the object is manufactured). We only sustain the configuration within which things can show themselves in this way, namely as hammers (or not). Within this configuration, it is the thing itself with its capacities that determines whether it does reveal itself in this way. If a thing is unsuitable for hammering, nothing we can say or do (except, of course, repairing or remaking it) will make it suitable. If a thing is suitable, it is so whether we notice it or not. Our practices at most determine the possibilities for what there can be, not what there is. This requirement is just as true for the objects of scientific research as it is for ordinary equipment like hammers. Whether something can be an oncogene that can serve as a probe for gene activity is determined by the context of practices within which DNA sequences, oncogenes, and genetic probes can be understood and dealt with. But whether a particular thing is an oncogene or a suitable probe will depend entirely upon the capacities of that thing. The context provides the necessary supporting conditions that enable these capacities (or rather, what we thereby can recognize as capacities) to become manifest even as capacities, and to be *significant.* Only then do they become determinations of a thing (or kind) as belonging to a kind or category of kinds. But when they do, which things possess those capacities becomes an entirely objective matter.

My third point further qualifies the apparently idealist implications of my initial claim. I have been speaking as if the configuration of social practices from which meaning emerges were abstractable from the things in the world that we encounter through those practices. But this image is misleading. Practices are not representations that can be understood abstractly. They are always ways of dealing *with* the world. The ontological kinds they make manifest are determinable only through our purposive interactions with things of those kinds, and thereby with the other things that surround us. And those other things are as essential to the existence of meaningful ontological possibilities as our practices are. There must be nails for there to be hammers; for there to be electrons, there must be such things as atoms, on the one hand, and cathode-ray tubes on the other. That is, there must be the things that they interact with and the equipment that enables us to interact with them.[50] Another way to put this is that for there to be things of any particular kinds, there must be a world to which they belong. But the reality of that world is not a hypothesis to be demonstrated; it is the already given condition that makes possible any meaningful action at all, including posing and demonstrating hypotheses.

Once we see this distinction, we can understand that the ontological categories "existent" and "nonexistent" ("real"/"unreal") are redundant in much the same way that "truth" and "falsity" are redundant. "Truth" adds nothing to the assertions to which it is applied. In a similar way, "existence" is not a real property; it adds nothing to the determinations of the things said to exist. The predicate "true" can be applied only to sentences in a language. So to this extent what is true (i.e., what *can* be true or false) depends upon what can be said in that language. But this constraint does not mean that the language has some determining interactions with the states of affairs to be spoken of. A language opens a field of possibilities for speaking (sentences that are true-or-false in Hacking's sense)[51] without in any way determining which are true and which are false. It connects assertions with truth conditions but does not determine whether those conditions obtain. Similarly, what exists depends upon the field of meaningful interaction and interpretation within which things can be encountered. This configuration of practices (including, of course, linguistic practice) allows things to show themselves as they are in a variety of respects. These showings are not independent of this configuration any more than what is true-or-false can be independent of a language. There cannot be things that cannot interact with the things disclosed within a meaningful world (as there cannot be truths or falsehoods that cannot be expressed in a language). Perhaps this point is obscured by the usual failure to recognize existence/nonexistence as a linked *pair* of terms comparable to true/false. Just as what is not a sentence in a language is not true-or-false, there is no fact of the matter about whether things that cannot intelligibly be encountered within a meaningful world exist or do not exist.

The recognition that the possible ways a thing can be depends upon

50. Heidegger 1957: 88; 1962: 121–22.
51. Hacking 1982.

the configuration of practices within which they become manifest should therefore not reinforce the realist's fear that we are being described as "world makers." The language we speak does not determine which of its sentences are true. The practices that constitute our "world" likewise do not determine which things exist, with what properties. Which things there are, what properties they have, and what relations they enter into are determined by the things themselves and "how things stand" with them. Existence and truth have similar relations to meaning. The conditions that must obtain for a sentence to be true constitute its meaning, and this depends upon the circumstances in which the sentence is appropriately used. What it is for an x to exist (as an x) is constituted in a similar way by the ways it can be encountered in the course of intelligibly dealing with the world. Both determinations are holistic. Understanding the meaning of a sentence requires understanding many other sentences along with it. Being able to recognize an x for what it is requires knowing how to recognize and deal with many other entities. An understanding of such appropriateness is sustained by the actual use of or other practical encounter with the things in question, in order to achieve the ends that call for such dealings.[52]

We can now see more clearly both the parallel and the essential difference between my view and those characteristic of linguistic empiricism. Both insist there is a connection between what counts as real and what can be manifest to us. Empiricists, however, believe that this connection rests upon some definite characteristics belonging to us (typically, the physical capabilities of our sensory apparatus). However contingent these capabilities may be, they circumscribe the limits of knowledge. Having taken the linguistic turn, they see these also as the limits to meaningful discourse; and since existence claims were thought to be meaningful only relative to a language, our sensory capabilities placed definite constraints upon what kinds of things there could ever (be said to) be. I see no grounds for any such ahistorical constraints. What can be manifest follows from the configuration of practices of the human community rather than from the physical characteristics of our sense organs.[53] This configuration is never fixed

52. Will (1981) develops an interesting argument parallel to mine, to the effect that a recognition of the epistemological importance of social practice is perfectly consistent with a commonsense realism.

53. I say *the* human community, because the different characteristics of different human communities must be to some extent adjudicable; the potential intelligibility of their practices, and hence of the things they encounter, is the condition of their counting as *Dasein;* of course this does not mean, pace Taylor, that we are always bound to take seriously *their self*-interpretation, but this is a matter for chapter 6.

once and for all but is potentially open to the manifestation of unimaginably strange and "unobservable" entities (perhaps even quarks!). The recognition that there are no a priori constraints on what there can be is the legacy of Heidegger's critique of essence and Quine's critique of analyticity. The result is a "pure" account of meaning in the sense in which Rorty called Davidson's a pure philosophy of language.[54] Learning "how meanings work" gives us no special insight into ontology or epistemology, for to say that knowledge, truth, and existence are practice bound and holistic tells us nothing about what those practices are (or must be) and what the relevant wholes consist in. The purity of these considerations of meaning therefore vindicates the realist's commonsense belief that the meanings that make sense of the world for us cannot impose any real determinations upon what there is.

The invocation of meaning as the arbiter of truth conditions for sentences and existence conditions for things may still seem to suggest the reintroduction of the notion of a conceptual scheme intervening between us and the world. But this interpretation would be a serious misunderstanding of what I have claimed. For my arguments about the redundancy of "existence" closely parallel Davidson's deconstruction of the notion of a conceptual scheme confronting an uninterpreted reality. Both of us deny the intelligibility of an uninterpreted reality and therefore of a scheme we impose upon it (though for Davidson the reasoning seems to run in the other direction). Davidson concludes that once the distinction of linguistic scheme and uninterpreted reality collapses, this leaves intact our ordinary notions of objectivity without requiring us to accept a realist notion of the "independence" of truth from our language. "Without the dogma, . . . truth of sentences remains relative to language, but that is as objective as can be. In giving up the dualism of scheme and world, we do not give up the world, but re-establish unmediated touch with the familiar objects whose antics make our sentences and opinions true or false."[55] The same is true of my treatment of existence and nonexistence. These notions remain relative to a configuration of practices in the same way that truth is relative to a language. There are only possible ways to be within the clearing opened by a configuration of practices. But this configuration is not a practical scheme confronting "independently" existing things; it is our unmediated dealings with things in the world, whose antics alone determine what there is and what determinations these things have.

54. Rorty 1979: 257–62.
55. Davidson 1984: 198.

Davidson, it is well known, develops his critique of the scheme/content dogma through a discussion of the principle of charity as an unavoidable constraint upon the translation of someone's utterances into our language. The unintelligibility of the idea of a conceptual scheme follows from the fact that translatability into our own language (however imperfect and underdetermined) is a necessary condition for emitted sounds or inscribed marks to count as an utterance in a language at all. A similar point can be made about the things that show up in our dealings with the world. Belonging to the realm of possible determinations open within our practices is constitutive of a thing's being a thing at all. But this claim is just to say that having determinate properties, and interacting with other things in ways we must take account of, is a necessary condition for a thing to be.

This point is difficult to recognize, because the things we ordinarily encounter, with their many evident properties, so clearly belong to the real (because of their many determined interconnections with other things) that it is hard to see the possibility of their thinghood's being in question. It is the same sort of difficulty we have in trying to hear or see sentences in a language we already know as meaningless sound or random marks on a page. What we need for clarification is an example where "thinghood" is in question. Fortunately, we have at hand a revealing case.

Consider the initial involvement of scientists with what eventually became recognized as thyrotropin releasing hormone (TRH). The name TRH was originally bestowed upon whatever was physiologically active (in this case stimulating the pituitary glands of rats to produce and release thyrotropin) in certain chromatographically isolated fractions of the hypothalami of sheep or pigs. But it was an open question whether TRH actually denoted a thing rather than an unstable artifact of the investigation. What was expected to answer this question was the successful (or unsuccessful) attribution of a chemical structure to TRH. That (real) substances have such structure, and what counts as having such structure, depends upon the complex of practices that had developed over a hundred years of biochemistry. Chemical structure in the modern sense is a way of being a substance that became a possibility only with the gradual introduction of this practical complex. But this development in no way determined what the structure of TRH was or even that there was a definite structural substance that constituted the physiologically active fragment of those fractions. It was the stuff in the fractions that ultimately revealed itself to be pyro-glu-his-pro-amide. But in doing so, it also revealed itself to be something at all. That TRH could be an artifact, a will-o'-the-wisp, as several other similarly isolated "releasing hormones" had turned out to be, remained at least until then a real possibility. And this is the crucial point: not to show up in the ways that allow something to count as an x (in this case, as a chemical substance) is not to be a thing at all.

The point is not that TRH magically became a real substance when its structure was successfully determined. At that point it was recognized as having been a substance all along, in large part because it had the appropriate sort of property (in this case a molecular structure). But this condition for being a substance was made possible historically by the evolving complex of practices that made molecular structures intelligible and allowed them to reveal themselves. Only within such a field of practices could having such a structure be a condition for counting as real. Of course no one doubted there was a powder at the bottom of a test tube that had resulted from the processing of millions of hypothalami. But many people quite reasonably doubted whether this powder represented a "substance" until it revealed itself to have the necessary determinations.

Perhaps I can best sum up my response to the realist by posing and answering a series of questions about this particular case. Is there (really) such a thing as TRH, a substance whose molecules are composed of a chain of three amino acids and an amide group? Yes. Does its existence mean anything more than that we take account of TRH in our dealings with the world? No. Could there be such a thing as TRH independent of certain practices and beliefs? No. Within the context of those practices and beliefs, is it in any way dependent upon us whether a particular sample of material is (or contains) TRH? No, except in the sense that we deliberately isolated this particular sample in order to obtain TRH. Is it true that TRH is pyro-glu-his-pro-amide? Yes. Does this identity mean anything more than that TRH is pyro-glu-his-pro-amide? No. Does the truth of this sentence explain why we can use TRH to stimulate the release of thyrotropin in rats? No. Does the structure of TRH itself (together with a great deal of physiology) explain this effect? Yes. Is the truth-or-falsity of "TRH is pyro-glu-his-pro-amide" (i.e., its being a meaningful candidate for truth or falsity, depending on whether some definite conditions obtain in the world) dependent upon our practices and the language we speak? Yes. Is the truth of this statement dependent upon our language, practices, or beliefs? No. Does "TRH" refer to TRH? Yes. Does this reference mean anything more than that it is recognized by competent practitioners that the phrase "TRH" is appropriately used in circumstances where a substance under discussion could in principle be identified as TRH? No.

When these points are recognized, my view no longer seems to put in doubt the truth or referential success of the best scientific theories, nor does it consign the pragmatic success of those theories to the realm of miracles. I happily endorse the pragmatist insight[56] that the world *is* what shows up in our practices. But I need not accept the metaphysical extravagances of convergent realism to do so.

56. See chapter 1 above, pp. 24–25.

Chapter 6

Natural Science, Human Science, and Political Criticism

We can now return to the Diltheyan distinction between the empiricist sciences of nature and the interpretive human sciences. This distinction has clearly collapsed in its traditional formulations, along with empiricist accounts of the natural sciences. But the recognition that even natural science is interpretive has led to a variety of attempts to distinguish the *ways* the two sorts of science are interpretive, or the *interests* at stake in the two sorts of interpretation. Furthermore, some of those philosophers who have been most sensitive to the differences between a theoretical and a practical hermeneutics have been in the forefront of attempts to resurrect the Diltheyan distinction.[1] The difference between the practical holism found in Heidegger and the sentential (or theoretical) holism of Quine and Davidson has been claimed to reflect the varying epistemological and political standing of the natural and human sciences.

These are not just idle disputes among philosophers. Dreyfus and Taylor first raise their concerns not in an abstract way, but through powerful, concrete criticism of predominant approaches to the study of psychology, political science, and sociology, which they take to be misguided attempts to imitate the methods of physics.[2] Habermas likewise draws upon long-standing criticism of "positivist" social science.[3] Their more general, neo-Diltheyan arguments are their at-

1. Dreyfus 1980; Taylor 1980.
2. Dreyfus 1979; Taylor 1964, 1979.
3. Habermas 1968a, 1971.

tempts to draw a moral from the specific criticisms of behaviorist and cognitive psychology and "value-neutral," empiricist social science.

For Taylor and Habermas, this distinction also has a political point to it. They claim that the study of human beings and their practices and institutions is always culturally situated and politically engaged. The ideal of a value-neutral human science is pernicious, for it masks its own uncritical assumption of particular values and political ideals and conceals the fact that different social realities are in part constituted by its own concepts and practices. The same is not true for the natural sciences. For Taylor, the natural sciences are different because they investigate the natural world we *all* share, regardless of our language, cultural background, or political projects, while the human sciences study practices *constituted* by language and social practice.[4] Thus, Taylor claims:

> We might say, for instance, that we have a vocabulary to describe the heavens that [the inhabitants of a traditional Japanese village] lack, namely, that of Newtonian mechanics; for here we assume that they live under the same heavens we do, only understand it differently. But it is not true that they have the same kind of bargaining as we do. The word, or whatever word of their language we translate as 'bargaining', must have an entirely different gloss, which is marked by the distinctions their vocabulary allows in contrast to those marked by ours. But this different gloss is not just a difference of vocabulary, but also one of social reality. But this still may be misleading as a way of putting the difference. For it might imply that there is a social reality which can be discovered in each society and which might exist quite independently of that society, or indeed of any vocabulary, as the heavens would exist whether men theorized about them or not. And this is not the case; the realities here are practices; and these cannot be identified in abstraction from the language we use to describe them, or invoke them, or carry them out.[5]

For Habermas, in contrast, what is distinctive of the natural sciences is not that they describe a common natural reality, but rather that they reflect our shared interest in the capacity to manipulate and control the natural world to satisfy our various ends. Because such capacities satisfy a universal human interest, the natural sciences are not open to political criticism in the same way as the human sciences, which serve varied political interests.[6] Thus, he argues against Marcuse:

4. Taylor 1979.
5. Taylor 1979: 44–45.
6. Habermas 1968b. The most relevant passages were translated into English as the appendix to the 1971 translation of 1968a.

> The idea of a New Science [in which the viewpoint of possible technical control would be replaced by one of preserving, fostering, and releasing the potentialities of nature] will not stand up to logical scrutiny any more than that of a New Technology, if indeed science is to retain the meaning of modern science inherently oriented to possible technical control. For this function, as for scientific-technical progress in general, there is no more "humane" substitute.[7]

These attempts to differentiate two kinds of science gain plausibility from even a cursory investigation of the contemporary standing of the natural sciences and the human sciences. The human sciences as a whole enjoy significantly less predictive success than do the natural sciences: they are less precise, less accurate, and their predictive capabilities are much more limited in scope. At the same time, they employ a less rich theoretical background: theory in the human sciences is more informal, more disputed, and less easily applicable to particular situations. Knowledge in the human sciences seems less cumulative, since fewer results achieve a standardized form that can be unproblematically applied in the course of further research. Furthermore, as imperfect as Kuhn's criteria for demarcating "normal science" may have been in many respects, they seem to demarcate further clear differences between human and natural science: in the human sciences, books (rather than articles in professional journals alone) are still important vehicles of scholarship, there is extensive controversy over fundamental issues of method and ontology, there is a lack of consensus over what counts as good work, and so on.[8] There are also at least de facto differences between the two kinds of science with respect to the types of criticism results are subjected to (the human sciences seem more open to political criticism) and the degree of their interaction with philosophy and literature.

These contrasts can easily be overdrawn, and it would certainly be a serious mistake to claim that the human sciences altogether lack predictive success, theoretical depth, or cumulative achievement. There are also exceptions to these broad patterns on both sides. But there seem to be real differences between the natural and human sciences in the ways they approach their research and the results they obtain. These differences cannot be easily dismissed, and Taylor, Dreyfus, Habermas, and others present a serious challenge when they try to explain them as reflecting intrinsic differences between two "natural kinds" of interpretation.

7. Habermas 1968b: 58, with interpolations from p. 55; English translation 1970: 88, with interpolations from p. 86.

8. Kuhn 1970a, chaps. 2–4.

The various neo-Diltheyan positions offer an additional attraction to many philosophers. At a time when the cultural significance of philosophy as the theory of knowledge has become more dubious, these positions suggest a distinctive substantive role for philosophy as the theory of interpretation, the discipline that notes the significance of and marks the differences between the various kinds of interpretation in our culture.[9] Philosophy becomes the hermeneutical counterpart to Husserl's eidetic theory of intentionality: in both cases, philosophy involves a turn from what becomes manifest in our experience to the phenomenon of manifestation itself and aims to mark off essential distinctions within it. These distinctions in turn have substantive, "foundational" implications for what can justifiably be said and done in the positive sciences.

One thrust of this book, first with its practical hermeneutics of natural science and then with its promised consideration of the political character of the study of nature, is to challenge these attempts to demonstrate intrinsic differences between two kinds of science. The former point tends to undercut Dreyfus's and Taylor's rationales for such a distinction, and the latter tends to challenge Taylor and Habermas. We must therefore confront in some detail the arguments that take these thinkers toward such different conclusions despite a similar account of and emphasis upon interpretive practices. My concern is oblique to theirs. I am not interested in challenging their interpretations of the human sciences, which are often persuasive in their details. I do intend to show that only a mistaken view of the natural sciences enables them to draw as sharp a distinction as they do between two different kinds of science, with allegedly different epistemological and political concerns.

There will be three stages to my argument. I will first challenge the specific arguments that have been used to establish some sort of fundamental difference between the interpretation of nature and the interpretation of human beings. I will then, in the second stage, consider why the scientific interpretation of nature cannot be so sharply distinguished from the study of human beings, at least not in the ways suggested by Dreyfus, Taylor, and Habermas. Here I will challenge the notion that our self-understanding is not at issue in our understanding of nature, and hence that we do not have as much at *stake* in the natural sciences, either politically or culturally, as we do in the study of human beings. As the final stage of my argument, I will

9. The most prominent recent critic of philosophy's *cultural* significance as the theory of knowledge has been Richard Rorty (1979, 1982).

suggest a different strategy for thinking about the differences between the natural and the human sciences, one that will return us to the theme of scientific knowledge as a mode of power. This strategy will also enable me to challenge yet another defense of such interpretive differences, this one formulated by Ian Hacking.

Two Kinds of Scientific Interpretation?

I shall begin by considering a family of arguments proposed by Taylor and Dreyfus, which aim to show that there are intrinsic methodological or ontological differences between the investigation of nature and the study of human beings. All these arguments rely to some extent upon a Heideggerian practical hermeneutics. Later in the chapter I shall turn to the somewhat different strategies employed by Habermas and by Hacking, which nevertheless lead to comparable conclusions.

There are at least four distinct arguments that Taylor and Dreyfus have developed to spell out and justify their central conclusion that there are essential differences between the natural and the human sciences. It will be useful to have them clearly enumerated at the outset. First, Taylor argues that although the natural sciences are interpretive, the human sciences are *doubly* interpretive. Not only are social scientists engaged in interpretation when they study human beings, but the objects of their study are self-interpreting, such that their self-interpretation must be taken into account in any adequate social scientific interpretation of them. This consideration, he thinks, must lead to differences in method, and more important, in the degree of objectivity obtainable in our interpretations and the predictive capability to be expected from them.[10]

In his second argument, Taylor claims that the same sorts of predicates cannot be applied to the objects of the two kinds of science. Natural science aims for what he calls "absoluteness"; this means that "the task of science is to give an account of the world as it is independently of the meanings it might have for human subjects, or of how it figures in their experience. An adequate scientific account should therefore eschew what one could call subject-related properties."[11] By contrast, Taylor thinks that subject-related properties are irreducible in the human sciences.

10. Taylor 1979, 1980.
11. Taylor 1980: 31.

> To understand someone is to understand his emotions, his aspirations, what he finds admirable and contemptible, what he loathes, what he yearns for, and so on. Understanding doesn't mean sharing these emotions, aspirations, loathings, etc., but it does mean seeing the point of them, seeing what is here which could be aspired to, loathed, etc. Seeing the point means grasping the objects concerned under their [subject-related] desirability–characterizations.[12]

This argument is a corollary to the first. In order to take seriously the self-interpretation of the *objects* of the human sciences, as the first argument required, we must describe them in terms that would be inappropriately "subjective" or anthropomorphic in the natural sciences. We do not have to see the point of molecular behavior in order to explain it—presumably because it does not *have* a point in the way human behavior does.

Dreyfus originated the third argument to show that the natural and human sciences are essentially different. He recognized that research in both the natural sciences and the human sciences requires the kind of tacit skills and holistic configuration of practices that he (and Heidegger) think cannot be adequately represented in a theory. But he argued that the natural sciences can arrive at interpretations that need make no reference to the skills and practices constituting them, while the human sciences face the impossible task of objectifying their own background practices. This difficulty arises because the practice of human science is itself a human activity to be scientifically comprehended, whereas the practice of natural science is not an object the natural sciences must account for.[13]

Dreyfus eventually rejected this argument, for reasons we shall discuss shortly. In doing so, however, he developed a fourth argument for essential difference. His latest claim is that there cannot be a *theory* of human capacities of the same sort as the theories of natural science, namely, one that is fully explicit, abstracted from particular cases, universally applicable, formulated in terms of discrete elements, systematic, complete, and predictive.[14] Such a theory is impossible because human capacities and activities depend upon tacit skills, which are articulated only through specific applications and presup-

12. Taylor 1980: 32.

13. Dreyfus 1980, 1984. He eventually abandoned this argument in 1984, admitting that human science need not after all objectively comprehend the conditions of its own possibility.

14. Dreyfus 1984: 8–14. He grants that there could be such a theory in principle but argues that in practice it could not be achieved in the only way we could reasonably expect—by abstracting elements from the everyday context of human activity.

pose a holistically understood context.[15] Furthermore, these skills must be employed just to identify successfully the *objects* of human science. These tacit skills are therefore indispensable: we cannot even replace them with a theoretical understanding that does not try to imitate what they can achieve.[16]

A common theme runs through all four of these arguments. Interpretation in the two kinds of science differs because in the human sciences our interpretations are more closely connected to our self-understanding, our language, and our practices. This connection in turn affects the methodology and the ontological standing of the interpretation. *We* are at issue in these interpretations, in a way we are not in the natural sciences.

The first part of my response to Dreyfus and Taylor is that the hermeneutical character of inquiry by itself carries no substantive implications for what the objects of the inquiry must be like. A Heideggerian hermeneutics does not justify the notion that human beings are special kinds of objects, whose interpretation requires special methods or serves distinct transcendental interests. We can see this objection more clearly by considering a logical difficulty with Taylor's and Dreyfus's arguments, first noticed by Mark Okrent.[17] Okrent argues that Taylor's and Dreyfus's arguments involve a paralogism comparable to the Paralogisms of the Soul in Kant's *Critique of Pure Reason.* Kant's Paralogisms make the point that the Transcendental Unity of Apperception, which he claims to be a necessary condition for the possibility of any experience, cannot itself be an object of experience and hence cannot be subsumed under the Categories, which presuppose the Transcendental Unity of Apperception for their application.[18] More generally, they show that we cannot treat the *subject* of experience, qua organizing and interpreting subject, to be itself an object of experience subsumable under the categories and rules applied by the subject in cognition. Okrent points out that in varying ways Taylor's and Dreyfus's arguments conflate essential characteristics of our interpretations qua meaningful interpretation with essential features of us, the interpreters, taken now as objects of an interpretation.

Consider specifically how Okrent's objection attaches to the arguments I have just summarized. The case is straightforward for Taylor's first argument, because the argument itself is a clear statement of what Okrent takes to be a paralogism. Taylor begins with the point I

15. Dreyfus 1984: 5–7.
16. Dreyfus 1984: 10–14.
17. Okrent 1984: 23–49.
18. Kant 1965: 135–64, 327–83 (A105–30, B129–49, A338–405, B398–432).

and his critics happily acknowledge: all knowledge is interpretive, and can be intelligible only against a background of meanings and practices. But he infers from this point that when we take ourselves or others as the objects of interpretation, we must take account of them *as* interpreters working within a particular configuration of meanings. As Okrent insists, this inference is a non sequitur. Scientists are obliged to conduct their studies of human beings within such a configuration of meaning, but this requirement says nothing about how human beings will show up as objects within that configuration. Even the most naive behaviorist psychology satisfies this condition, as does indeed any program of investigation of which we can make any sense at all. The hermeneutical account of understanding applies universally and does not distinguish one discipline or theory from another.

Taylor's argument nevertheless acquires some apparent plausibility from an ambiguity as to what sort of argument it is, and consequently an ambiguity concerning the basis for Taylor's conclusions. I have taken it so far as if he were proposing a general constraint upon any investigation of human beings as acting or interacting (for Taylor does not propose hermeneutic constraints upon human biology or biochemistry). Taylor often speaks this way—for example, when he argues that exact prediction in the human sciences is "radically impossible."[19] But in addition to the paralogism ascribed by Okrent, there is something further that is deeply incoherent in Taylor's argument when it is construed this way. The thrust of hermeneuticist arguments like Taylor's is holistic and anti-essentialist. What things are, and what characteristics they can have, depends in part upon the practical configuration within which they become manifest. There are no essences independent of this configuration of practices and the language invoked within it. Therefore, to argue that there are essential constraints on a particular type of science (human science) based upon an essential characteristic of all human beings (being self-interpretive or self-defining) does not seem consonant with his approach.

Fortunately there is another reading of Taylor's argument that escapes both the paralogism and this related hermeneutical incoherence, though at the cost of eliminating any *in principle* differences between human and natural science. Instead of an argument from outside social scientific discourse, concerning necessary constraints upon any possible investigation of human beings acting or interacting, Taylor may intend to challenge mainstream social science from within

19. Taylor 1979: 68–71.

its own concerns. Taylor encourages such an interpretation by juxtaposing to his general, methodological argument an attempt to show what gets left out of mainstream political science and how this impoverishes its understanding of political life. What is left out, of course, is this practical configuration within which the more manifest activities studied by political scientists make sense. The difficulty Taylor recognizes in his line of attack is that the practices and underlying assumptions of mainstream political science tend to make the phenomena he is pointing toward invisible. So the response to his arguments by social scientists is likely to be perplexity over what he is talking about and an inclination to dismiss these phenomena as insignificant if they exist at all. Anticipating this resistance, Taylor tries to show that these phenomena, which he claims are not amenable to analysis by mainstream social science, are nevertheless causally efficacious in politically significant ways. Phenomena that are clearly manifest even to hidebound empiricist political scientists, such as the development of the counterculture and the New Left in the 1960s, cannot be adequately explained without invoking the sort of hermeneutic "methods" he advocates.[20]

It is in this sort of argument that Taylor is at his best, and although it needs much more extensive development and indeed must become a full-fledged research program to be actually acceptable on these grounds, I find the argument promisingly persuasive.[21] But the prospects for a Taylorian political science are not what is at issue here. What is important for my purposes is that this reading reduces to insignificance the import of Taylor's argument for an essential difference between the natural sciences and the study of human beings. Let us suppose that Taylor is right, so that in at least some significant cases, treatments of human behavior that attribute causal efficacy to changes in people's self-description turn out to be epistemically superior to accounts that do not make such attributions. This concession would satisfy his more limited purpose of reforming current practice in social science, but it would in no way reestablish a distinction between two epistemologically distinct sorts of science, for two reasons. The first reason is that Taylor's claim would have to be established by arguments within each particular scientific field, arguments that would not be markedly different epistemologically from their counterparts in the natural sciences. Questions about which sorts of entities, processes, and so on, should be taken to be causally efficacious

20. Taylor 1979: 55–65.

21. This may well represent a difference between Okrent and me concerning the outcome of these arguments, since he seems more optimistic about the prospects for something like mainstream empiricist social science.

are internal to every discipline and are resolved by appealing to the same sorts of criteria—for example, fruitfulness, simplicity, breadth of scope. Taylor would have presented not a general philosophical argument but a series of particular arguments for the various fields of human science. The arguments would have to be established independently, for reasons internal to each discipline, and would be independently defeasible. They would appeal not to essential features of inquiry, but to the particular circumstances of each discipline at a given time in its development, circumstances that are of course open to change.

The second reason follows up on the first. Different fields of science quite naturally take account of different kinds of entities and processes in explaining events in their domains. Biologists refer to differential selective pressures on populations of organisms, or the imperfect transmission of information in molecular transcription in explaining some biologically significant events. There may be no obvious analogues in physics to such explanations, but we do not conclude from this lack that biological science is essentially different from physical science or that inanimate and animate objects require different methods of study. Nor should we be surprised if the human sciences need to utilize a type of description that has no place in biological or physical science, but this need by itself would give us no warrant for postulating essential differences between two types of science or between natural and social "worlds." There is only one world, and overlapping aspects of it are studied by different disciplines with their own characteristic methods and problems.

These responses to Taylor's first argument that the natural and human sciences are essentially different can be easily adapted to apply both to the second argument (which Taylor also developed) and to the third (the first one offered by Dreyfus). Taylor's second argument, for example, concerns the use of "subject-related" predicates in social scientific discourse. If Taylor is claiming that the role of interpretation in human understanding necessitates this approach, he commits the paralogism, for as Okrent nicely summarizes the complaint, "It still does not follow from the supposed fact that X understands *himself* by using concept Y, that a social scientist must understand X by using Y. To argue that it does follow is once again to confuse a condition for understanding on the part of a person who understands with a *real* predicate of that person under a different description."[22] The self-understanding of X and the social scientist's

22. Okrent (1984), deleted from published version.

understanding of X are both interpretations intelligible within a local configuration of practices and goals, but there is no a priori reason these configurations must overlap or coincide.[23]

If, on the other hand, Taylor is claiming only that we will produce a more adequate political science (or psychology, economics, etc.) by utilizing subject-related properties, then the argument must be established in each case, with reference to the goals and practices of each discipline. It constitutes not an appeal to essential differences between types of science and regions of objects, but simply a pragmatic attempt to redirect a particular field of inquiry from within.

The third argument, developed by Dreyfus, is somewhat more complicated. He is not claiming, as is Taylor, that an empiricist methodology in social science is peculiarly inappropriate to its subject matter, and that a hermeneutic method would be more satisfactory. Dreyfus draws the much stronger conclusion that *no* methodological strategy can ever be fully satisfactory in social science. Like Taylor, Dreyfus appeals to the need to comprehend the background practices that allow for the intelligibility of human thought and action. But he focuses upon the fact that the social sciences themselves involve practices that are cases of human behavior and therefore prime candidates for social scientific understanding. These practices elude self-comprehension, he thinks. Some approaches to social science, such as ethnomethodology, may focus upon the practices of their competitors, but they cannot do the same with their own investigations. Ultimately, Dreyfus concluded, any attempt to develop a Kuhnian "normal science" in the human sciences should give rise to "deviant" approaches that attempt to thematize the background practices that make the "normal" approach intelligible. Thus the difference alleged to exist between natural and social or human science is not that there are essential differences between the methodologies or ontologies appropriate in each case, but that the social sciences must always be afflicted with methodological and ontological instability. The point is

23. It may sometimes be desirable to be able to show why agents understand themselves in the way they do, or why they fail to understand themselves in the terms of the social scientist, but this does not require that social scientists *adopt* this description as a feature of their own interpretation. It may not even require them to *understand* the agent's self-interpretation in its own terms. Their interpretation of the agents' behavior must meet some kind of coherence test and presumably must account for the agents' speech behavior, but this does not imply that acceptability of that interpretation *by the agent* is a criterion for a successful social scientific interpretation. There are in any case well-known difficulties with the very notion of understanding another agent in the way the agent understands himself or herself, since the only criteria we have for assessing this still depend upon the *interpreter's* understanding rather than the agent's.

not that the natural sciences do not involve similar background practices, but that they need not take account of these practices as objects of investigation. The study of physics is not a physical phenomenon to be explained by physics in the way that sociology is a phenomenon for sociological study or that cognitive psychology might want or need to explain its own psychological cognition.[24]

My response to this argument is twofold. First, the expectation that the social sciences account for their own background practices invokes a variant of the paralogism described by Okrent. It is illegitimate to demand this task of them. Second, however, there is a sense in which the background practices of a science are a legitimate issue "within" that science, but *this* sort of epistemological reflexiveness is equally characteristic of the natural sciences.

The first point should by now be straightforward. Dreyfus's argument acquires its force only from the assumption that the background practices of a science must be comprehended *as* a meaningful configuration of practices. For the argument is not that social scientists can say nothing about their own practices, but that they cannot explain what accounts for the *intelligibility* of those practices. If behaviorist psychology had worked, for example, it might well have been able to predict the *motions* of researchers in behaviorist psychology. What it would have had to leave out, according to Dreyfus, are the concerns that gave a point to *doing* behaviorist psychology. But the argument purporting to show that it needs to account for these concerns is the Okrent paralogism: because human beings must engage in interpretive practices in order to study anything (including behaviorist psychology), any interpretation that takes human beings as its objects must describe them as self-interpreting beings. And this requirement, I hope it is clear by now, is a non sequitur. As I noted earlier, Dreyfus himself has come to recognize the difficulty and has abandoned this argument.[25]

Having said all this, I should add that there is a legitimate concern with their own background practices in the human sciences, but this concern is just as important in physics or biology. To see this parallel, consider ethnomethodology. Dreyfus had once seen ethnomethodology, with its reflexive interest in the data-gathering practices of the social sciences, as a paradigm case of the epistemological reflexivity involved in studying human beings. This type of scientific reflection might show, for example, through a study of coroners' practices in compiling suicide statistics, that Durkheim's celebrated correlation of

24. Dreyfus 1984.
25. Dreyfus 1984: 7–8.

suicide and anomie is an artifact of the ways the data were originally classified.[26] But this concern is not unique to the social sciences; a physicist or biologist may be very interested in whether certain results are an artifact of the investigation. This was certainly true in the investigation of TRH we discussed earlier. Everything done in the field hinged on whether the apparently stable traces of physiological activity shown by various hypothalamic fractions were the effect of a "substance" or were an artifact of the assay procedure and its interpretation. An important theme of Guillemin's influential review article of 1963[27] was that this question could not be answered in a satisfactory way using current approaches to the field and that therefore a more ambitious program of study was called for. More generally, we saw throughout chapter 4 that scientists are constantly concerned with the sensitivity, accuracy, fruitfulness, and competence of their practices. These practices are not themselves *objects* of the science in the way Dreyfus had claimed that social science must be an object of study for itself. But scientists' concern to understand their objects of study leads them to be explicitly concerned with their practices of investigation. Precisely the sort of epistemological dilemmas that such reflexiveness is supposed to lead to[28] crop up repeatedly in natural science, whenever scientists are forced to examine critically the ways their objects of study become manifest to them. The problem of adjudicating between alternative readings of the same situation has never been unique to the study of human beings. A more traditional Diltheyan might argue that understanding coroners' practices requires us to decipher conscious (or unconscious) intentions on the part of Catholic and Protestant coroners, whereas substantiating TRH does not. This conclusion is undoubtedly true, but it is not clear what follows from it, since further argument would be required to show that intentions pose special problems of interpretation. Indeed, one motivation for Dreyfus's and Taylor's approaches to the issue was to avoid the problems encountered in trying to base their distinction upon the peculiarities of interpreting the mental.

Even if this objection were sound, so that we can no longer claim there are essential ontological or methodological differences between the human and natural sciences owing to the self-interpretive character of human beings, there is still the fourth argument proposed by Dreyfus. He now claims that the particular project of the natural

26. Douglas 1967.
27. Guillemin 1963.
28. Taylor 1979.

sciences since the seventeenth century can be understood in terms of a notion of "theory" that, while not impossible in principle to develop for the human sciences, poses practically insuperable obstacles to its extension beyond the study of nature. Dreyfus draws upon a group of Heidegger's later essays[29] to claim that the modern natural sciences aim to fulfill the Platonic ideal of *theoria.* This ideal requires that a "theory" must be explicitly formulable in full, must be independent of its application to any particular case, must range over discrete elements, and must be systematic (aiming ideally toward a consistent, unified theory that leaves nothing out). Dreyfus then argues that such theories are highly unlikely to be achieved in the study of human thought and interaction, despite their demonstrable power and importance in the natural sciences. His argument refers once more to the tacit skills and holistically understood background that his Heideggerian hermeneutics ascribes to human activities and capacities. But the argument now turns to the practices the human sciences study rather than the practices of those sciences themselves. Dreyfus's claim is that the human sciences cannot both have *theories* in the strong sense described above and also use those theories to predict events in familiar terms. His example, drawn from Pierre Bourdieu, concerns Lévi-Strauss's account of gift giving. What counts as a gift (rather than an exchange, say) is determined by the background of practices of a particular society. For reasons discussed in chapter 3 above, Dreyfus thinks that scientists cannot replace the tacit grasp of that background with a formal set of rules capable of generating the same recognition of gifts. But such a set of rules, operating upon decontextualized elements, is precisely what Dreyfus thinks a *theory* modeled on natural scientific theories must provide.[30] He therefore concludes:

> The problem is that the social theorists do not and cannot have a theory of how the people they are studying determine what counts as a gift. The problem is that the social theorists' way of defining a gift, since it must be abstracted from the everyday world, will not coincide with the pragmatic way the social situation defines a gift. There is no corresponding problem for bubble chambers, since bubble chambers do not classify their own bubble-tracks.[31]

There are two problems with this argument. The first is the familiar paralogism. Dreyfus sees this problem and recognizes that he has

29. Dreyfus 1981, 1984; Heidegger 1952, 1954, 1962.
30. Dreyfus 1981, 1984.
31. Dreyfus 1984: 12.

not shown the *impossibility* of a theory of human capacities. But he argues that the possibility of a predictive, theoretical social science is of little practical importance, because we have no idea how to begin to produce the relevant sorts of theory. Dreyfus claims we cannot appeal to our everyday understanding of social life in order to discover the appropriate theoretical vocabulary, because the features we understand in this way cannot be comprehended in a decontextualized way.[32] Thus, he says:

> It follows from the above considerations that the right vocabulary for what were once called the human sciences would have to be a vocabulary which picked out entirely different features than those abstracted from our everyday activity—features or attributes which would remain invariant in different pragmatic situations and across cultural revolutions. Just what such features would be, no one can say.[33]

Such a theory thus remains unattainable in practice. It also may be undesirable even if attainable, because what we would most like to predict in social life would be events under their everyday descriptions: how people will vote, when they will give gifts, what they will purchase at a given price, and so forth. If their behavior was predictable in terms that were not straightforwardly translatable into such everyday terms, the predictions might not be particularly useful to us.

Even if this claim was true, however, Dreyfus's argument would not establish the difference he postulates between social and natural science, because his argument depends upon a mistaken conception of theory in the natural sciences. The discussion of science in chapters 2 and 4 undermines all five of Dreyfus's constraints on what counts as an adequate theory. Theories cannot be fully explicit; instead, they embody possibilities for further interpretation of the objects in their domain (this potential is what McMullin described as the metaphorical character of scientific theory). Theories are also not abstracted from particular examples, as we saw in discussing Cartwright's account of models and Kuhn's account of paradigms. Scientific theories are focused on paradigmatic examples of their "application," and the content of the theory cannot be fully understood in abstraction from them. The elements of theory are not discrete and context free, but are picked out only against a background of research practices. Nor are theories systematic in Dreyfus's sense: their unity is established by practical skills rather than by formal rules, and incompatible theories may be appropriate in the same domain for different purposes.

32. Dreyfus 1984: 10.
33. Dreyfus 1984: 14.

Thus the fourth argument to show that there are essential methodological or ontological differences between the natural and the human sciences also fails, because it presupposes a conception of (natural) scientific theory that my earlier arguments have shown is seriously flawed. Dreyfus may seem to have a stronger point when he discusses the degree of completeness and predictability attainable in the two domains of science. Later in the chapter, however, I will show that this feature has little to do with the types of theories we employ and much to do with the sorts of facilities we have and the manipulations we are willing and able to perform in order to establish predictable situations in these two domains of science.

The Political Importance of the Interpretation of Nature

The arguments presented so far have been attempts to undermine the claim that human beings, as interpretive subjects, present special interpretive difficulties to any science that would take them as its objects. But these arguments justify only a limited verdict on Taylor's and Dreyfus's claims: Not proved. Perhaps they or someone else could construct a better argument that would avoid the paralogism and *would* establish some essential difference between the interpretation of nature and the interpretation of human beings and their practices and institutions. In this part of the chapter—which constitutes the second stage of my argument against the idea that the natural and human sciences are essentially different—I will show why better arguments for such a distinction are not likely to be forthcoming.

The underlying intuition that seems to motivate Taylor and Dreyfus to argue anew for their claim is that there is something at stake in our scientific self-understanding (i.e., human science) that is not at stake in our understanding of nature. Our self-understanding matters to us because, as self-interpreting beings, we are constituted by it. Nature, they suggest, is not (self-)interpreting and hence simply is what it is regardless of how we interpret it to be. In an important sense, then, Taylor and Dreyfus are realists about the natural world but constructivists about the ontology of the social world.

This underlying intuition mistakes the place of the understanding of nature in our self-understanding, and it does so in part by failing to grasp the implications of Heidegger's treatment of the hermeneutic circle, even though his account is an important influence on both Taylor's and Dreyfus's views. We cannot take for granted the conception of nature that underlies their neo-Diltheyan project. The manifestation of nature as dead, inert, noncommunicative, and fit only for manipulation to suit our ends belongs to our configuration of practices, not to a "real," uninterpreted nature of things. The point is not that we ascribe crudely anthropomorphic characteristics to nature or that nature is our "construct." We encounter "nature" through our practices, as it fits in and is revealed intelligibly in that context. What it is to *be* natural is at issue in our dealings with the world (we must not forget that distinctions like natural/social, natural/artificial, natural/deviant articulate a field of meanings). It is not that different cultures live in different worlds, but that the world has a different hold upon them.

The importance of recognizing that the natural world exists and is manifested socially and historically is emphasized when we grasp the extent to which our *self*-interpretation is at stake in our understanding of nature. This involvement can be seen in a variety of ways: in the interaction between our social or psychological categories and the processes and structures in terms of which we interpret the natural world (including both the anthropomorphizing of nature and the naturalizing of self and society); in the social/psychological constitution of "objectivity" and of nature as the correlate of our being objective; in the forms of agency through which we engage nature as we understand it; and above all, in the ways the interpretation of nature meshes with the ways power operates within our lives.

It will be useful to begin to show this connectedness with a point about Heidegger and the hermeneutic circle, since this discussion will lead us naturally into an account of the significance of the interpretation of nature. We have seen how Dreyfus and Taylor want to use the claim that human beings are self-interpreting; they argue that the human sciences are always interpretations of an interpretation and cannot achieve the stability and closure possible in interpreting nature, which is not itself interpretive. But in developing the point in this way, they fail to grasp what is new and decisive in Heidegger's hermeneutics. Traditional Diltheyan hermeneutics emphasized the meaningful character of the *object* of interpretation, and the hermeneutic circle involved an interplay between the object as a meaningful whole and the parts that both compose the whole and acquire their sense from it. For Heidegger, the hermeneutic circle is an interplay between the *understanding* of the world as the meaningful configuration within which things are manifest as what they are and the *interpretation* of particular things within the world. The circle thus has the same structure for the interpretation of persons and of things, because it has nothing to do with the presumptively meaningful character of the object.[34]

34. Okrent, n.d., appendix to part 1.

Dreyfus's and Taylor's claims may nevertheless seem supported by Heidegger's account when he says that we (Dasein) understand ourselves in terms of possibilities and when he concludes that "only Dasein can be meaningful or meaningless."[35] The natural world, after all, does not understand itself in terms of possibilities, and if only Dasein is meaningful, surely that must make some difference in how we come to interpret it. But this is to misread Heidegger in a fundamental way. Understanding themselves in terms of possibilities is not a real property of the beings whose way of being is Dasein; it is a condition on the possibility of there being an understanding of being (i.e., any disclosure of beings) at all. The possibilities in terms of which Dasein understands itself are not something distinct from its understanding of the world; they *are* its understanding of the world, including "nature." To understand oneself in terms of possibilities is to understand the world as a field of possible (self-interpretive) action. This is the configuration within which anything becomes intelligible, not just Dasein. To say with Heidegger that only Dasein is meaningful is not to say that only human beings "have" meaning, but rather to say that a practical, purposive configuration or world is the condition for anything's having any intelligible properties of any sort. Meaning is a "formal" condition[36] on the intelligibility of beings rather than a substantive characteristic of some particular being.

Put this way, the point is abstract and difficult to grasp. It becomes clearer when we consider its implications for understanding the natural world. Dreyfus and Taylor think that our language and practices constitute us as the beings we are, because who we are is at issue in those practices. In the natural sciences, by contrast, they speak of language and practice as merely enabling the world to reveal itself as it "already" is. Heidegger's analysis (as I extended it in chapter 5) reveals that this is mistaken on three counts. First, it is not only "social reality"[37] that somehow depends upon language and practice; the natural world likewise acquires a definite character only within a purposive configuration of practices, because this configuration determines what can count as a thing, property, or relation.[38] Second, who we are is just as much at issue in the natural sciences as in those inquiries that make us directly into an object of study. This means that the political aspects of self-interpretation that so concern Taylor and Habermas (and Dreyfus when the "political" issue is understood as nihilism, i.e., the absence of any political issues as itself an issue) are

35. Heidegger 1957: 151, 1962: 193.
36. Heidegger calls it a "formal-existential framework" (ibid.).
37. This is Taylor's phrase (1979).
38. See above, chapter 5, pp. 152–65.

crucial to the natural sciences as well as to the human sciences. This point will be my central concern for the rest of the book, which returns explicitly to the question of the relation between knowledge and power. Third, to say that who *we* are is at issue in both the human and the natural sciences may suggest a certain privileged position for us. Dreyfus seizes upon this implication to suggest at one point that who we are is at issue not in the sciences themselves, but in the history, philosophy, and sociology of science, where we are once again the direct object of scientific inquiry.[39] But to accept this suggestion is to be misled by Dreyfus's otherwise useful colloquial formulation of the issue as one of "who we are." Who we are can be worked out only through an understanding of the world, and strictly speaking it is not we (as particular beings) who are at issue in our interpretations, but what it is to be. This is what Heidegger called the "meaning of being" in *Being and Time* and what I have called the "configuration" of things (and practices and purposes) that give us a hold upon the world.[40]

The claim that our self-interpretation is at stake in the natural sciences also—and that this issue involves them inescapably in the politics of interpretation—is the most fundamental reason why Dreyfus's and Taylor's distinction cannot be sustained. It is important enough, and controversial enough, that the arguments above in terms of Heidegger's hermeneutics[41] must be supplemented by a further analysis that gives some concrete sense to the claim that there is at stake in the natural sciences a political issue concerning what it means to be.

It may be useful to give some initial characterization of what I mean here by "political." I begin with Taylor's criticism of attempts to confine political science to a discussion of actions that can be described in "brute data terms." Taylor claimed that if we can describe only particular actions, and neglect the Heideggerian configuration of language and practice within which those actions take place, we will fail to understand even those actions that are explicitly and recognizably political, such as voting or demonstrating. For our self-understanding as agents, and as members of a community in which only certain sorts of actions and social relations are intelligible, shapes our political possibilities and on occasion can provide the focus of political conflict. I shall argue here that our scientific practices, and the way the natural world shows up through those practices, are an important part of that

39. Dreyfus 1980.

40. Chapter 3 above, pp. 63–64.

41. Okrent's claim that these arguments involve a paralogism also implicitly draws upon Heidegger.

configuration of language and practice. Scientific practices belong to what Foucault has called the realm of "government":

> "Government" did not refer only to political structures or to the management of states, . . . the legitimately constituted forms of political or economic subjection, but also modes of action, more or less considered and calculated, which were destined to act upon the possibilities of action of other people. To govern, in this sense, is to structure the possible field of action of others.[42]

A field of action is constituted both by material surroundings and technical capabilities and by the shared understanding of what it makes sense to do and to be in those surroundings. Scientific practice is political in the sense that it helps structure our field of possible action in both ways. It transforms our material surroundings and technical capabilities; it also helps shape (and is shaped by) the concepts and practices that make action intelligible. It would be a mistake to think of scientific research as political in the way that elections, rebellions, and legislation are political. Science is political in its influence upon (and its being influenced by) the practical configuration within which politics as narrowly defined takes place.

The political character of scientific interpretation can show itself at both the macro and the micro levels. To begin with the former, consider the seventeenth-century developments that brought the classical sciences of astronomy and mechanics into something like their modern form. Everyone grants that this revolution fundamentally changed the understanding of the natural world. But it did not just substitute one position for another in response to a common issue; it changed what physics was *about*, what was at issue in understanding the natural world. In so doing, it also fundamentally changed what it was to be human. I do not propose to retell this familiar story here; I shall only point to certain features of the story that bear emphasis in the light of what I have already argued.

That the scientific revolution changed what physics was about and what it aimed for should no longer be controversial. It introduced new "ideals of natural order,"[43] stripped places of their physical significance (i.e., dissolved the distinction between sub- and superlunary spheres and removed the structure of center and periphery, and of natural places and directions, from the world),[44] burst the closure of

42. Foucault 1983: 221.
43. Toulmin 1961.
44. Kuhn 1957.

the world to reveal an infinite universe,[45] and substituted the categories of force, mass, and instantaneous velocity for qualitative and teleological ones in the comprehension of matter.[46] This revolution changed what was at issue in being human. The Cartesian subject of modern philosophy emerged from reflection on how such a physical order could be comprehended and known; the relation between human beings and the divine took on new shapes; the problem emerged of whether and how physical and moral descriptions of persons were compatible; and both the gendering of nature and the sexual imagery of our knowledge of it were revised.[47] These were not only intellectual changes. They were connected to transformations in political relations and institutions, the creation of new religious practices, and the emergence of new forms of economic behavior and social interaction.

Dreyfus and Taylor seem inclined to say that in the case of the changes in physics, our conceptions of the physical world changed without changing the world itself, whereas in the case of the corresponding changes in our understanding of self, God, society, and the good, the world itself was transformed. For reasons that should now be familiar (both from chaps. 3, 4, and 5 and from much of the history and philosophy of science since Koyre and Kuhn), I see these changes to be all of a piece. In each case the world as we are dealing with it has changed, and it resists or accommodates our activities in novel ways. The world, even the natural world, is always a field of activity, and the things we can encounter within it acquire definiteness through the possible dealings we can have with them.

Dreyfus's claim that we encounter the natural world theoretically as present-at-hand was an attempt to recognize this primacy of practice in disclosing the world while evading the consequences I am now ascribing to it. His claim was, in effect, that though everything acquires definite characteristics only within a practical context, the objects of natural science acquire in this way the characteristic of being decontextualized. I have already argued in chapter 4 that this claim involves a misunderstanding of science as a practice. What we can now see is that it also involves a misunderstanding of nature and its place within the configuration of practices that includes modern science. Nature has not been exempted from the configuring of a field of activity but has been made subordinate within it. For us moderns

45. Koyre 1957.
46. Kuhn 1957.
47. Keller 1985; Merchant 1980.

the natural world has become not neutral, but plastic, able to be utilized and exploited to a variety of ends. Heidegger attempted to capture this aspect of our world by saying that everything becomes *Bestand*, or resource.[48] Others have described the same phenomenon as the death or domination of nature.[49] These claims will have to be treated more extensively and critically in chapter 7, when we focus upon the relation between knowledge and power. But we should be prepared to realize now that the issue of our relation to the natural world (and indeed, the issue embedded in the question whether there is a "natural world" distinguishable from the world as we encounter it through our social practices) must be taken as a political issue in the broad sense. We differ from other societies and cultures precisely on whether nature is to be responded to as sacred or as personified, where this latter point is not simply the presumably settled question of whether there "is" a mind or soul in nature but rather the question whether nature has some sort of moral standing (e.g., being worthy of "respect"). Our understanding of this issue fundamentally shapes what we do and how we deal with one another. It is at least as central to and definitive of our world as is the "bargaining culture" Taylor sees as central to our political self-understanding.[50] To say that this issue has already been decided by science, which shows the natural world to be merely present-at-hand, is to beg the question.

There is another side to the recognition that our relation to the natural world may involve political issues, for our understanding of nature as object also invokes the ideal of objectivity in assessing our attempts to know nature. This ideal is in turn deeply bound up with our self-conception as knowers and as agents. As Heidegger has pointed out, objectivity is a characteristic of the subject in the guise of a revealing of the object. Objectivity is motivated by the concern to let the *object* show itself as it is, untainted by our idiosyncratic constructions and interpretations. But this becomes a concern to *view* the object in the right way, that is, to adopt the appropriate methods and posture *as a subject*.[51] Which aspects of ourselves as knowers are essential to rightly (objectively) viewing the world and which ones are idiosyncratic, anthropomorphic, or "subjective" is at issue in our understanding of nature. Not all conceptions of objectivity are the same. The modern conception, for instance, insists on divorcing subject and object, on excluding all affects as antithetical to objective understanding, and on

48. Heidegger 1954.
49. Merchant 1980; Leiss 1974.
50. Taylor 1979.
51. Heidegger 1952.

taking manipulation and control as reflecting only objective capabilities and not subjective needs or interests. Moreover, this conception is itself open to criticism. One might possibly use, for example, a psychoanalytic model to claim that commonplace modern interpretations of objectivity reflect inadequate (e.g., immature and/or androcentric) forms of psychological development or a sociopolitical criticism of objectivity as an ideology of domination.[52] But in any case, the point is that what would be an appropriate stance for understanding nature objectively cannot be separated from the political questions of who we are and how we can and should relate to one another.

The political character of the scientific conceptions of nature can be further clarified by considering some of the explicitly political uses to which these conceptions have been put. For the line between what is and what is not part of the given, noninterpretive "natural" world has not simply paralleled the one between the human and the nonhuman (even if this distinction could be given a clear formulation, which I doubt). The distinction has been used to render gender distinctions and roles "natural" and thus exempt from social redefinition, and even to parallel the male/female distinction to that between the socially self-defining and what is given and "natural." Comparable interpretations have been given to the relation between Western and "primitive" cultures and races, though these are perhaps now more readily discredited. "Nature" is a thoroughly political concept and marks a difference in how we deal with things and people, and what sort of standing they have, rather than a given distinction between two kinds of reality.

The political character of interpretation in the natural sciences is not limited to this sort of broad, sweeping distinction between the natural and the human/social, however. Much of chapter 7 will be devoted to articulating the political significance of the kinds of local scientific practices and competences described in chapter 4. One example here will perhaps illustrate the point that everyday natural scientific practice and knowledge share with the study of human beings an involvement with the categories and practices of our self-understanding and thus an involvement in our politics broadly construed.

The example I have in mind is a certain kind of causal story common in biology and biochemistry, in which complex events within a cell or a multicellular system are claimed to be controlled or regulated by a molecule(s) of a specific substance. Such causal explanations have

52. Keller 1985; Leiss 1974.

been appropriately called "master molecule" accounts.[53] A classic example of this is the so-called central dogma of molecular genetics, according to which the production of proteins within a cell was thought to be governed by a one-way transfer of information from DNA to RNA to protein. The postulated role of releasing hormones (such as TRH, which we discussed earlier) in regulating the secretion of hormones by the pituitary invokes the same sort of explanation. To see the importance of such an example, we must keep in mind the importance of causal explanation in science, which was perhaps once masked by positivist scruples but has been increasingly recognized, particularly within the context of what I have been calling the new empiricism.[54] Causal explanations are more stringent than mere correlations, and the difference is scientifically important precisely because an understanding in causal terms provides a basis for experimental (and of course technological) intervention.

In fact, the connection between causal explanation and human agency is two-way. On the one hand, the significance of causal explanation instead of mere correlation derives from its relation to possible intervention by manipulating the relevant causes. On the other hand, our conception of causal connection depends upon how we understand agency.[55] The counterfactual import of causal claims is justified by the possibility of our intervening to make the cause of an event present where it was not before, or to remove it. Von Wright notes:

> The past is closed, but the future, we said, is open. This implies that, although we cannot interfere with the past and make it different from what it *was*, we may be able to interfere with the future and make it different from what it otherwise *would be*. It is on this possibility, viz., of interfering with the "normal" course of nature, that the possibility of distinguishing the nomic from the accidental ultimately rests. . . . [Thus] the distinction between accidental and nomic regularities is based on the idea of making experimental interferences with nature.[56]

This interpretation suggests that our understanding of human agency and our invocation of causal laws in the natural sciences have to be understood as interdependent.

The example of "master molecule" explanations can also lead us to

53. Nanney 1957, cited in Keller 1982.

54. See Cartwright 1983, esp. essays 1 and 4, and Hacking 1983, chap. 2.

55. A classic version of this argument can be found in G. H. von Wright (1974).

56. Wright 1974: 38–39, 120. It should be emphasized that it is the *idea* of causality itself, and not the particular case of causal efficacy, that is dependent upon our understanding of human agency and the scientist's role in experimental intervention.

reflect that different conceptions of agency may yield different understandings of causation and that this can make a difference in our understanding of the sciences. There is a political metaphor built into such explanations. A one-way causal relationship invokes the notion of complete control or regulation by an authority impervious to influence or interference. But a very different political model is built into interactionist accounts of cellular functioning. As one geneticist described such alternative accounts:

> The first of these [concepts of genetic mechanisms] we will designate as the "Master Molecule" concept, . . . in essence the Theory of the Gene, interpreted to suggest a totalitarian government. . . . By [the second concept, the "Steady State"] . . . we envision a dynamic self-perpetuating organization of a variety of molecular species which owes its specific properties not to the characteristic of any one kind of molecule, but to the functional interrelationships of these molecular species.[57]

Our understanding of the concepts involved here ("regulation," "organization," etc.) probably cannot be entirely separated from their political content.

But the point goes deeper than just the claim that political ideologies have perhaps influenced our scientific descriptions, in ways that, after all, we might be able to eliminate if we were more careful. For the kinds of causal explanations we invoke are closely connected to the kinds of experimental interventions we engage in. Scientists study the cellular systems involved in the expression of genes by intervening in them in ways that lend themselves to one-way causal interpretation. They initiate changes in cellular DNA (e.g., by X-ray mutation or controlled recombination) and track a specific outcome (production of a particular protein, appearance of a phenotypic trait, etc.). Similarly, they might inject a solution containing a releasing hormone into an animal and measure the rate at which the animal secretes the substance the releasing hormone is supposed to regulate. Scientists' own agency in investigating the system has the same causal character they attribute to the master molecules, and the one is particularly well suited to reveal the other. Such experiments do not aim to uncover other effects of the changes introduced, nor do they investigate directly the complex relationships through which such simple regulation is effected. One should not, of course, grossly oversimplify this argument to claim that experimental intervention *cannot* reveal

57. Nanney 1957: 136, cited in Keller 1982: 154.

complex functional interactions or that scientists are not sensitive to the possibility of such complexity. Nor should this be taken as an epistemological objection to such experiments. These interpretations would reduce the argument to silliness. The point is only that our own power relations with the materials we investigate, and our conception of our own agency as scientists, may more readily reveal those materials in the same causal terms and may *incline* us toward such causal explanation. Different kinds of experimental practice and different self-conceptions might well incline us in a different explanatory direction.

We are seeing here in the natural sciences something analogous to Taylor's objection to speaking of a given "social reality." I have already cited Taylor's concern that this locution

> might imply that there is a social reality which can be discovered in each society and which might exist quite independently of the vocabulary of that society, or indeed of any vocabulary, as the heavens would exist whether men theorized about them or not. And this is not the case; the realities here are practices; and these cannot be identified in abstraction from the language we use to describe them, or invoke them, or carry them out.[58]

Now we have learned from Heidegger that the crucial factor is not that the things we are describing *are* practices; it is that the relations of *disclosure* through which entities are *manifest* to us are practices. And we have seen more concretely that the things disclosed through these practices, even the heavens or the functioning of a cell, "cannot be identified in abstraction from the language we use to describe [these practices], or invoke them, or carry them out."

Cognitive and Political Interests in Science

I have been arguing, contrary to Taylor and Dreyfus, that our self-understanding and the practices through which we sustain and develop it are at stake in the interpretation of nature, as well as in the explicitly human sciences. We cannot take natural reality for granted while focusing upon the social world as the locus of political controversy. These arguments undercut Taylor's and Dreyfus's reason for believing there must be an essential difference between two distinct kinds of science. This objection can now be reinforced by showing

58. Taylor 1979: 45.

that the considerations I have just been raising make untenable two further attempts to demonstrate such an essential difference. Both of these attempts focus explicitly upon the question of what is politically at issue in the two kinds of science. The first was another approach to the same issue by Dreyfus. The second was proposed by Jürgen Habermas in some of his earlier writings.[59] The criticism of these two proposals will complete the second stage of my argument in this chapter.

This new argument by Dreyfus does not aim to show that the human sciences cannot be like the natural sciences. He argues instead that if they were made to imitate the natural sciences, they would be nihilistic in a way that an objective natural science is not. Such a model of the human sciences would require that scientists decontextualize their self-understanding and treat it as a field of objects for description and analysis. To objectify our self-understanding and treat it as a fully describable system of objects (e.g., beliefs, desires, and values) would loosen any hold that understanding might have upon us and therefore leave our possible choices of self-interpretation bereft of any public significance. Nothing would matter to us. A decontextualized understanding of our choices and interaction would deprive them of their point. This loss would be a cultural disaster that would have no parallel in the outcome of a successful natural science, because its objects do not *have* a point to lose.[60]

The arguments in the preceding section show the difficulty in Dreyfus's attempt to exempt the natural sciences from any direct contribution to the nihilism he finds in contemporary culture. Dreyfus characterizes nihilism as the loss of any serious public issues at stake in our social practices. Issues would no longer be serious, because we would understand ourselves as subjects whose beliefs and values can be objectified, clarified, and chosen. When all our concerns are values we can freely choose or reject, then there is no issue on which the choices between them can seriously matter.[61] But how is this nihilism supposed to bear upon the relation between the human and natural sciences? Dreyfus thinks a fully objectified human science would be nihilistic because it would objectify our *self*-understanding, including all matters of serious concern to us. An objectified natural

59. Habermas 1968a,b, 1970, 1971.

60. Dreyfus 1980, secs. 6–7, 1981.

61. Dreyfus 1980, 1981. One might also compare Kierkegaard's version of this claim (1954: 203), comparing the self-objectifying subject to a monarch in a country in which revolution is legitimate.

science would be acceptable, however, because it would objectify and subject to our will only the *things* around us. But to sustain this distinction, one must make a deep assumption: what can be seriously at issue in our self-understanding has nothing to do with our understanding of nature and our relation to nature. For if it did, then the objectification of nature would also involve a nihilistic objectification of our self-understanding. I have, of course, been arguing that we cannot separate natural science from our understanding of ourselves. Dreyfus himself sees this at one point. When he speaks of possible sources of resistance to the encroachments of nihilism, he turns to the still extant fragments of a "pretechnological understanding of the meaning of Being":

> Now scattered in our inherent background practices, this understanding involves nonobjectifying and nonsubjectifying ways of relating to nature, material objects and human beings. . . . To take examples close to home, Faulkner personifies the wilderness, Pirsig speaks with respect of the quality even of technological things such as motorcycles, and Melville opposes Ishmael as mortal and preserver to Ahab as the willful mobilizer of all beings toward his arbitrary ends.[62]

These examples illustrate how our understanding of nature is crucial to our grasp of what is at stake in our lives. If Dreyfus is right that nihilism is the greatest cultural danger confronting us, then his willingness to take nature and science for granted may itself contribute to this danger.

Habermas's argument for essential differences between the human and natural sciences in some ways reverses the approach taken by both Taylor and Dreyfus. They argued that the interpretive field opened by our practices and purposes is socially constructed and historically situated, and that this feature renders the human sciences irreducibly unstable, nonpredictive, or both. They thought the natural sciences are free of the marks of social construction because they have been decontextualized from this field of interpretive practice.[63] Habermas, by contrast, rejects the idea that there can be a decontextualized, disinterested scientific cognition of nature as present-at-hand. The natural sciences, like all cognitive practices, are guided by particular human interests. He argues, however, that some of these interests are not merely socially and historically constructed, but uni-

62. Dreyfus 1980: 22.
63. Dreyfus 1980, 1981; Taylor 1979, 1980.

versal. The natural sciences reflect our interest in the technical control of the things around us. Because this interest is universal, Habermas claims that the technical capabilities provided by the natural sciences are thereby immune to political criticism. Such is not the case with the universal interest in communication and consensus that governs our interpretations of one another (including the human sciences). This interest compels us to take up and focus upon the political issues involved in our self-definition, for it cannot be satisfied without taking seriously the way our communications and practices project political goals. Thus, ironically, Dreyfus and Taylor, and Habermas, argue for a similar distinction between two kinds of science, for opposing reasons. The former argue that all interests and purposes are shaped by a historical context of social practices, but that natural science is a decontextualizing practice. The latter argues that natural science cannot be disinterested or decontextualized, but he thinks that it satisfies a universal, ahistorical human interest that in the end distinguishes it from human science in a similar way.

The natural sciences, according to Habermas, satisfy our interest in the successful manipulation and control of things around us. Their aim is technically exploitable knowledge that can fulfill our material needs and desires (there is thus an internal, cognitive connection between natural science and technology on Habermas's view). The human sciences, by contrast, are "subject to a constitutive interest in the preservation and expansion of the intersubjectivity of possible action-orienting mutual understanding. The understanding of meaning is directed in its very structure toward the attainment of possible consensus among actors in the framework of a self-understanding derived from tradition."[64] These different interests reflect the difference between two fundamental human projects, which Habermas calls "work" and "symbolic interaction."

Habermas thinks this difference in interests and projects manifests itself throughout the methods and results of the sciences. Technology, and the empirical sciences of nature associated with it, share the characteristics of work: for example, they are guided by technical rules rather than social norms, they employ context-free language, and their failure is marked by inefficacy rather than punishment.[65]

64. Habermas 1968b: 158, translated in Habermas 1971: 310.

65. Habermas (1968b: 64, 1970: 93) also describes other characteristics of work: acquisition by learning skills, orientation toward solving definite problems, conditional rather than reciprocal prediction, and growth of technical capabilities as the "rationalized" aim; however, only those mentioned in the text will matter to our subsequent criticism of Habermas's distinction.

This last difference is perhaps the most revealing. Failure to satisfy our technical interests in controlling our environment leads to ineffective or incompetent behavior, for which the *sanctions* against failure are "built in." In contrast, failure to satisfy our interest in communication and consensus leads to deviant behavior, against which sanctions must be applied if any adverse consequences are to occur at all.[66]

This difference is reflected in a fundamental political difference between the two kinds of science. Political interpretation and criticism are at the heart of the human sciences, for their aim is an understanding that can guide action and help coordinate our actions with those of others. There is, however, little scope for such criticism in the natural sciences, which concern not the orientation of action but the acquisition of the means for its successful completion. No matter how much our political interaction changes our attitudes toward nature or our theoretical understanding of how it behaves, we cannot do without something fundamentally like our technologies and the technical-scientific practices that have been developed with them.

> Technological development follows a logic that corresponds to the structure of purposive-rational action regulated by its own results, which is in fact the structure of *work*. Realizing this, it is impossible to envisage how, as long as the organization of human nature does not change and as long therefore as we have to achieve self-preservation through social labor and with the aid of means that substitute for work, we could renounce technology, more particularly *our* technology, in favor of a qualitatively different one.[67]

Thus the natural sciences develop perennial possibilities for controlling nature and serving our ends, while the human sciences must reflect our changing self-interpretations and goals.

There are three fundamental problems with this account as I see it. First, Habermas's conception of "work" that underlies this argument makes the same sort of error as Dreyfus's conception of "theory" I criticized earlier, by failing to grasp the practical hermeneutical character of science and technology. Second, Habermas's argument presupposes the identification of ends or purposes of action with social goals separable from their involvement with nature and treats nature solely as a neutral, exploitable means to human ends. This identity may well be what is at issue in a political criticism of science or technology, and it cannot be presupposed at the outset without begging

66. Habermas 1968b: 62–63, 1970: 92–93.
67. Habermas 1968b: 56–57, 1970: 87.

the question. Finally, in order to universalize the interest in technical control, Habermas must make it so abstract as to pass over many of the politically interesting questions. It may make a crucial difference which things are to be subjected to technical control, and in what such control for the sake of self-preservation would consist. To take a sharply drawn example, such a universal interest could presumably be satisfied either by adapting *ourselves* and our practices to minimize their ecological impact or by a massive substitution of technical systems for those natural processes not amenable to precise technical control. Yet the difference between these two approaches might well be of interest to critics of the technical employment of scientific knowledge.

Consider the first problem, concerning the differences between work and symbolic interaction. The claim that work (and the sciences and technologies that contribute to it) is based upon context-free language and explicit technical rules has difficulties that should be clear from our discussion in chapters 2–4. Scientific language and practice are embedded in a socially regulated and tacitly understood context. Scientific understanding cannot be articulated in the form of explicit rules any more than it can be formulated as decontextualized theory. The other claim, that technical failure is different from social sanction, may perhaps seem more plausible until we reflect that what counts as technical success or failure is socially negotiated and is enforced by social sanction (largely internalized by practitioners who share the goals of their community). Success and failure are shaped by the configuration of practices within which science becomes an intelligible activity, and they cannot be so sharply separated from the "symbolic interaction," "social norms," and punitive sanctions of a community of practitioners.[68]

The second problem concerns Habermas's attempt to separate means from ends and to relegate the natural world without question to the realm of means. We cannot simply dismiss the claim that natural science itself involves a kind of symbolic interaction with what we study. Certainly there have been scientists who in reflecting upon their work insist upon the importance of this kind of interaction, of "a feeling for the organism,"[69] even of love.[70] Perhaps this is metaphor or romanticizing, but it cannot be dismissed without argument or with the assumption that "nature cannot be regarded as a partner in

68. Chapter 4 above, pp. 92–95, 119–25.
69. Keller 1985.
70. Goodfield 1981: 228–36.

dialogue."[71] Furthermore, our dealings with the natural world may well be a focus of symbolic interaction. We understand ourselves in part as natural beings, and we certainly interpret ourselves through our dealings with other things in the natural world. It is difficult to confine the symbolic interaction through which we interpret ourselves to a limited sphere of dealings with one another, which exclude our interpretation of the (natural) world within which we find ourselves. Finally, if we take Heidegger's central points seriously (see chap. 3), it becomes difficult to characterize a rationalized realm of means distinct from the purposive configuration of practices within which they show up. It is only in the context of some issue(s) at stake in our practices that anything shows up meaningfully at all, including the objects of natural science and the material for technical manipulation and control.

The final problem for Habermas can also be put briefly, partly because it is simple, and partly because we shall discuss the relevant issues at greater length in the next chapter. Habermas attempts to isolate our technical capabilities for manipulation and control of things from political criticism on the grounds that they satisfy a universal human interest. Even if this assertion were right, however, it would leave still at issue what kinds of manipulation are appropriate and for which things this treatment would be appropriate. For if this truly were a universal interest always satisfied whenever human communities continue to exist, then this interest and its universality could not provide the basis for an argument in favor of the particular scientific and technological practices of our society. There have been alternatives to our science and technology, and these were presumably also ways of satisfying these universal interests. Habermas could try to argue that our science is not fundamentally different from others in its project, only in its efficacy. But this approach would require specific and detailed argument that he does not provide and would still leave open the issue whether this degree of efficacy is itself desirable.

Thus Habermas's attempt to insulate the natural sciences from political criticism is beset with serious problems. Its failure reinforces what my response to Taylor and Dreyfus showed, namely, that we have the same kind of stake in our scientific understanding of nature as we do in the human sciences. The claim that there is an essential methodological or ontological difference between the natural and the human sciences is most implausible. We are left with the conclusion that the understanding of nature embodied in our scientific and tech-

71. Hesse 1980: 181.

nical practices is integral to our self-understanding, in all the interpretive political complexity that Taylor in particular has so powerfully described. This conclusion completes the second stage of my argument in this chapter.

An Alternative View of Natural and Human Science

There may nevertheless remain nagging doubts about the result of these arguments, because however difficult it may be to provide a credible account of an ontological difference between social and natural reality that would compel a difference between the sciences that study them, there still seem to be persistent and important differences between the two kinds of science. If this discrepancy is not to be accounted for in terms of different kinds of interpretation, as Dreyfus and Taylor suggest, or of different interests at stake in the interpretation, as Habermas has proposed, then some other account may still seem called for. Rorty's suggestion that the differential successes of natural science are just coincidence and may be ephemeral seems unconvincing.

The basis for such an alternative account is available in what we have already said about the natural sciences. For we can plausibly explain the more prominent differences in style and results between the natural and human sciences by appealing to the practices of experimentally constructing and manipulating laboratory microworlds and to the interplay between these forms of intervention and theoretical modeling. Perhaps, as Bruno Latour once suggested, the difference between science and other practices is not one of methods, norms or standards, "culture," theoretical language, or the like, but rather the material difference embodied in laboratories and laboratory practices.[72] Moreover, this difference might then itself be explicated in terms of differences in the forms of power articulated and deployed in the disciplines.

To see this possibility, consider briefly some other attempts to account for the apparent differences between human and natural science, namely, the relative complexity of social phenomena, the causal influence of social scientific practice upon social behavior, and the good fortune of the natural sciences to stumble upon the "right" vocabulary for effective prediction and control.[73] It is not evident that natural events and processes are any less complex than social ones,

72. Latour 1983: 160–69.
73. For someone who makes the latter claim, cf. Rorty (1979: 351–53).

and it is not clear how one could even begin to evaluate this claim. What is the measure of complexity that would allow us to compare on a common scale the biochemical processes in a cell with the voting behavior of a political community? No such measure is needed, however, to recognize that the construction of phenomena in laboratory microworlds aims to reduce the complexity of the events under study and to introduce a comprehensible order into the situation. By artificially isolating and simplifying complex situations and introducing controlled disturbances, scientists come to understand events in their laboratory better than they can understand events outside. By modeling their understanding of outside events on their laboratory understanding, and perhaps even more by introducing laboratory equipment, materials, and procedures into the world "outside" in order to simplify and regulate it, scientists gain insights and capabilities that would have been inaccessible without the laboratory. There is of course a long history, especially in psychology, of attempting to develop a comparable experimental science of human behavior or cognition. But a little reflection upon the importance of carefully controlled and purified materials for successful experimentation will emphasize the crucial differences in most cases. Where are the human science analogues to chemical stocks, cell cultures, and laboratory-bred animals? We cannot (and I hope would not want to) produce "laboratory people" or laboratory societies that would enable us to intervene in artificially constructed human microworlds in controlled and trackable ways. Nor do I find attractive the prospect of introducing more predictable people into the world in order to *make* social life more predictable and manipulable.

A related argument can be made about the claim that human science cannot be stable and predictive because it changes the very practices it attempts to comprehend. This objection is also true of the natural sciences, which as we noted have created many of the phenomena they enable us to understand most clearly[74] and which have physically transformed the world in myriad ways. But these changes were to some extent deliberately and carefully introduced as limited extensions of laboratory capabilities. Innovations in the social sciences seem to have been disseminated in less controlled ways, although this deserves careful study that it cannot be given here.

It might be objected that we are talking about different things here—laboratory practices have physically rearranged the world, while the human sciences have primarily influenced the behavior and

74. Chapters 1 and 4 above.

the self-conception of the people they have studied. But it is not clear that this distinction represents anything more than particular differences between the influences of particular sciences. The development of artificial immunization and pharmacology has affected the ecology and ultimately the genetic development of certain viruses and bacteria, whereas the physical sciences have not usually had this sort of effect upon the things they characteristically study. This contrast has not led anyone to argue that the biological and physical sciences are essentially different from one another epistemologically. But why should, for example, the genetic development of penicillin-resistant bacteria be less significant epistemologically than the use of "rational expectations" to reduce the impact of anti-inflationary policies? In both cases the manipulations that both shaped scientific knowledge and were made possible by it have changed the world in which they originally operated and forced us to qualify or modify our original scientific claims. It could also be argued that the natural sciences have usually been content to understand the world *as changed* by its practices, whereas the human sciences (Marx notwithstanding) aim to understand human behavior as it was before the attempt to understand it. Dreyfus suggests something similar in saying that the human sciences must account for behavior like gift giving in its original terms, whereas the natural sciences are freer to reinterpret their objects to achieve scientific ends. This distinction may also have been what Habermas was trying to capture in saying that the two kinds of science reflect different interests. If so, however, it remains to be argued that this represents more than a contingent characteristic of science as we have practiced it.

Finally, if the natural sciences have managed to discover the "right" vocabulary for prediction and control (or at least an efficacious one), this achievement is not accidental. An important point of the practice of creating phenomena is to allow certain properties and relations to stand out clearly as significant and as causally efficacious. If the vocabulary of the science that results reflects a concern to understand just those properties and relations that are most clearly manifest in the *phenomena* (in Hacking's specific sense)[75] it has available for study, we should hardly be surprised. The modern scientific revolution was led by the study of some of the few naturally occurring phenomena—the movements of the stars and planets against the night sky—and it coincided with the development of more systematic and precise inves-

75. Hacking 1983, chap. 13.

tigation of these phenomena.[76] Subsequent developments, including the opening of new domains of investigation, were strongly influenced by the creation of new phenomena within those domains.

Consider, for example, the role of techniques for the confinement, identification, and weighing of gases in the development of eighteenth-century chemistry, or the sequence of technical developments that have been so central to the rapid progress of genetics over the past eighty years: Morgan's group's development of the *Drosophila* system (including the cytological techniques for studying chromosomes and the use of X rays to induce mutations), the turn to bacteria and bacteriophages as preferred organisms to investigate, the opening of molecular structures to investigation in the 1950s and 1960s, and finally the use of recombinant techniques to intervene directly at the molecular level. The growth of scientific knowledge is substantially focused upon the growth of techniques for isolating, intervening in, and tracking phenomena. These capacities of manipulation and control—or, as I shall characterize them in the next chapter, these tactics of power—are central to the de facto differences between the natural and the human sciences. Latour and Woolgar give a useful parody of the differences in material and technical resources between the two kinds of investigation, differences that reflect the presence or absence of these sorts of tactics:

> Occasionally, when members of the [neuroendocrinology] laboratory derided the relative weakness and fragility of the [sociological] observer's data, the observer pointed out the extent of the imbalance between the resources which the two parties enjoyed. "In order to redress this imbalance, we would require about a hundred observers of this one setting, each with the same power over their subjects as you have over your animals. In other words, we should have TV monitoring in each office; we should be able to bug the phones and the desks; we should have complete freedom to take EEG's; and we would reserve the right to chop off participants' heads when internal examination was necessary. With this kind of freedom, we could produce hard data."[77]

One need not endorse the efficacy or desirability of their observer's suggestions to recognize the differences between the natural and

76. The phenomena of the movements of the stars and planets were of course always available; it was the increasingly careful measurement and tracking of those movements, most notably by Tycho Brahe but also by others before him, that helped set the stage for the transformation of astronomy.

77. Latour and Woolgar 1979: 256–57.

human sciences with respect to the power relations between scientists and what they study. The point is not that social scientists could produce the same capabilities if only they were free to deploy the same power. There is still the question of designing good experiments, and before that the issue of developing useful materials, techniques, and phenomena. Most laboratory experiments, particularly of a new sort, result in ambiguous or even unintelligible outcomes, and most instruments and procedures produce noise rather than clear data. One can say of laboratory experiments what Arthur Fine once said about scientific realism: "The problem . . . is how to explain the *occasional success* of a strategy that *usually fails.*"[78] But even if the resources and techniques of laboratory science are not sufficient by themselves to produce the kinds of results that have distinguished the natural sciences at their best, they may well be necessary. It is certainly difficult to imagine the modern natural sciences without them.

The proponents of interpretive social science might object that this discussion is all beside the point, however.[79] They never thought that the social sciences are or should be interested in manipulation and control. It is not surprising that the natural sciences are better at this, because the human sciences are not playing the same game. Laboratories would then be beside the point; the real issue is the essential difference in the kind of interpretation involved, the same essential difference I thought I had ruled out by my earlier arguments. My response to this proposal is that even if they are right, the relevant difference is in the *power* relations the various fields of science involve. My argument that the forms of manipulation and control developed in laboratories must be understood in terms of power will be developed in chapter 7. But an interpretive social science would also develop a kind of power/knowledge. This aspect of the human sciences has frequently been noted in critical studies of anthropological or psychoanalytic interpretation. It has been carefully argued for in Foucault's study *The History of Sexuality,* which I shall also discuss in chapter 7.[80] So my claim is not simply that the natural sciences have laboratories, which the human sciences may not want or need, but that they deploy more effective forms of power. This claim does not reinvoke the sort of distinction I argued against earlier.

Another objection may also arise at this point, however. If we take seriously the claim that the apparent differences between the natural

78. Fine 1986: 119.
79. I am grateful to Brian Fay for pointing out this possible objection.
80. Foucault 1978.

and human sciences reflect material differences in the kinds of facilities and resources and the techniques of manipulation and control available to the two sorts of science, then a slightly different approach to accounting for these differences may lead us back to something like Dreyfus's and Taylor's versions. Such an approach was recently suggested by Ian Hacking. We have already seen an affinity between Hacking's contribution to a new empiricist philosophy of science and the Heideggerian practical hermeneutics I have been attempting to develop. He now argues that although there is a sense in which both the natural and the human sciences construct the objects of their investigation, there is a significant difference in the ontological status of the objects thereby constructed. This difference is reminiscent of Taylor's insistence that language constitutes social reality but only describes a preexisting natural world. Hacking argues that the creation of phenomena in the laboratory differs fundamentally from the ways new social practices and linguistic categories create new kinds of people (or new ways of being people). Physical phenomena are created historically, under idiosyncratic local conditions, but are not historically "constituted" because they are remarkably resilient to changes in the context in which they arose.[81] "The objects of the physical sciences are largely created by people, [but] once created, there is no reason except human backsliding why they should not continue to persist."[82] This is not the case with social phenomena, whose existence is rather more dependent upon particular social circumstances. For Hacking the difference is ontological: ultimately, a "dynamic nominalism" may be needed to interpret social reality, but a hard-headed "experimental realism" is called for concerning the things turned up by the natural sciences.[83] "We remake the world, but we make up people."[84]

I want not so much to collapse Hacking's distinction altogether as to blur it. When all is said and done, people and their practices and institutions may be more malleable than the world in which they find themselves. But the distinction is not nearly so sharp or obvious as Hacking (or Taylor and Dreyfus) seems to think. There are two reasons for this lack of definition. One is that the phenomena of experimental science are somewhat more historically variant than Hacking seems to think. The other is that there is a parallel in the human sciences to Hacking's experimental realism, one easily overlooked because Hacking makes a subtle shift in his level of description when he

81. Hacking 1984: 114–24.
82. Hacking 1984: 119.
83. Hacking 1984: 118, 122.
84. Hacking 1984: 124.

moves from the natural to the human sciences. Hacking claims to be a realist about the objects of natural science but an anti-realist about theories describing them. A comparable stance may be implicit in his views about the human sciences.

With respect to the first reason, the phenomena that have been central to the natural sciences could not have been created unless certain technical skills, facilities, and materials were available. These factors are not simply cumulative, for craft skills disappear and facilities and materials become unavailable (Hacking himself recognizes this attrition and cites the examples of brass working and lens grinding as disappearing skills). This loss of expertise may also make invisible those differences, distinctions, and connections that were discernible only by the skilled worker. Nor are these only technical skills ancillary to science. We are no longer able to reproduce what alchemists did in their laboratories (to some extent because we can no longer quite see the point of their manipulations and productions). More contemporary capabilities are also vulnerable. For example, with the rise of molecular genetics and the corresponding shifts in the interests of researchers, some of the skills and discernments essential to classical chromosome studies may gradually wither. How many scientists today are familiar enough with the chromosomal structure of maize to reproduce or even recognize McClintock's original evidence for genetic transposition? Likewise, many instrumental skills must gradually disappear as they lose any standard place in laboratory practice.

Perhaps deeper than the loss of skills and materials, however, is the possibility that some phenomena may cease to interest us, or may even lose their point, as the focus and direction of scientific work change. The creation of phenomena, after all, belongs to specific configurations of scientific practice. The point of a particular experiment, instrument, or technique can change; it can also disappear. Hacking sees the point that research is a practice with changing resources and interests, so that phenomena may in fact become obsolete, insignificant, or irreproducible. Why, then, does he dismiss this loss as only the result of "backsliding"? I think it can only be because he sees *phenomena* as decontextualized in Dreyfus's sense and hence as cumulative without regard to changes in the configuration of practices in which they arose. The phenomena persist even though the context in which they were originally produced *and understood* may have changed. Hacking illustrates his claim effectively with an example drawn from Kuhn's more recent work: "Kuhn writes that Volta saw his invention on analogy with the Leyden jar. Volta's description of it is strange, and we cannot credit his drawings, for they build in the wrong analogies. (When we examine them closely we want to say that the cells cannot have been made like that, for they simply cannot work.) But they worked. Current did flow. Once that had been done, physics never looked back."[85] I think Hacking is mistaking the standardization of phenomena (and instruments and materials) for their decontextualization, just as we have seen Dreyfus do.[86] What was once a fragile and very specialized achievement becomes robust and ordinary, adaptable to a variety of situations and materials. But this change does not mean the achievement has become independent of any social or practical context; it has only received a standardized form, less specifically tailored to particular locations or uses. Perhaps more charitably, we might attribute to Hacking the view that the possibility of standardizing experimental equipment and results is itself sufficient to remove from the objects of natural science the historical variability that is irreducible in us as people. We do not know how to produce standardized people.

The discussion of standardization in chapter 4 suggests, however, that such contextual variation is reduced in the natural sciences in this way, but not eliminated. We can produce voltaic cells and electrical currents in multiple circumstances and despite a significantly changed understanding of the chemical and electrical processes involved. But that does not mean that we could produce the phenomenon in just *any* circumstances. Electrical current and electrical theory still depend upon many aspects of practices like ours, ranging from the mining and working of metals to their classification and interpretation, or from the purposes for which making and having batteries and currents would be useful to the presence of other equipment with which they function. Our world would admittedly have to change dramatically for something so pervasive as the voltaic cell to become unintelligible or uninteresting, but it would require an untoward identification with our ways of dealing with the world to insist that such changes could be described only as "backsliding."

We can also blur Hacking's distinction in a second way. He argues that in the natural sciences the *entities* we deal with are already there, the *phenomena* through which they show themselves are created by us but then represent perennial possibilities, and the *theories* that describe and explain the entities and phenomena are variable and instrumental.[87] In the human sciences, however, he thinks that the

85. Hacking 1984: 118, with the parenthetical remark interpolated from 116.
86. I discussed Dreyfus's version of this mistake in chapter 4 above, pp. 112–18.
87. Hacking 1983.

descriptions and categories we use to describe people also in some sense *constitute* them as the people they are. *We* are not "real," preexisting entities. "Categories of people come into existence at the same time as kinds of people come into being to fit those categories, and there is a two-way interaction between these processes."[88] But now we are talking about the level of *description,* about which Hacking is not a realist even in the natural sciences. He is a realist only about the entities we are trying to describe. One might well therefore be such a realist about human *bodies* (embodied persons), for these we also remake rather than "make up." Then the categories through which we describe these bodies as persons would be comparable to the theoretical categories through which we interpret the behavior of physical things. Hacking's dynamic nominalism would be appropriate in both cases at the level of categorization. There would thus still be a level at which we could identify the same *things* under discussion in two very different categorizations. It is no objection to this approach to say that "body" is also an interpretive classification. So is "electron." The point in both cases is not that there is a privileged or true description of these things or their modes of action, but that the objects can be identified (according to some causal theory of reference)[89] independent of any particular description of them. The appropriate conclusion for both people *and* things would be, "We remake them rather than making them up, but we make up the categories through which they show up in the first place, as things to deal with, respond to, and remake."

Conclusion: Science and Political Criticism

This response to Hacking completes the final stage of my arguments concerning the parallels or differences between the natural and human sciences. What conclusions can be drawn from this entire complex of arguments? I have been challenging attempts to establish a general ontological or epistemological difference between two basic kinds of science, which appeal to supposedly different kinds of interpretation, theory, and interests involved in practicing science. The aim (and I hope the result) of my arguments is not to amalgamate the natural and social sciences or to reform them to reflect a "real similarity" supposedly uncovered here. They are no more essentially the same than they are essentially different. There are, of course, many

88. Hacking 1984: 122.
89. Hacking 1983, chap. 6.

specific differences of style and content between particular disciplines, and I have argued in the preceding section that there are techniques of power that may be largely confined to the natural sciences. We may even do well to preserve some or all of these differences.

What we cannot legitimately do is to take these differences as given or natural. This constraint has been the principal point of this chapter. We must examine and criticize all the sciences as part of our overall effort to understand ourselves, our world, the possibilities it offers us and what is at stake in them. Natural science, technology, and the nature revealed by them are integral to the world we inhabit. Nature exists in a social milieu and shows up in definite ways in response to specific social practices. At the same time, these practices help shape a field of intelligible possibilities for acting, and thereby for interpreting ourselves. To exempt natural science and the nature it reveals from such examination, or to take them as given, would be to court epistemological and political failure. And this danger, it seems to me, is what is problematic in the positions taken by Taylor, Dreyfus, Habermas, and Hacking. They are open to recognizing that the human sciences might embody techniques, strategies, or ideologies of power, domination, or meaning-dissolution and that these call for critical discussion. But they exempt the natural sciences from the same criticism. Or rather, they attempt to provide an unquestioned political legitimation for the practices of the natural sciences and require that we take them as the given context within which all social self-examination must take place.

There is an understandable reason for this attitude, which I initially considered in my opening chapter. That reason is the belief that power and domination were, at least in the contexts in which we are discussing knowledge, to be equated with the suppression or the ideological distortion of knowledge. Knowledge itself would therefore be justifiably exempt from criticism, while the social forces that threatened to impinge upon it would not be. To open knowledge, and the sciences that are our foremost examples of the pursuit of knowledge, to political criticism would be to aid and abet the forces of power and domination in the effort to suppress "inconvenient" or "dangerous" truths.

We have seen, however, that we cannot simply dismiss power as something repressive or distorting, from which we must isolate the ivory tower of truth seeking. There is a sense in which power is productive of knowledge and at the same time shapes the configuration of practices, aims, and equipment through which we understand ourselves and orient our activities. Power in this sense permeates the natural sciences and raises some of the same kinds of political issues about our self-understanding that Taylor, Habermas, and the others find in the human sciences. My criticism of these arguments concerning the differences between the study of humanity and the study of nature has therefore aimed to clear the way for a critical understanding of the interaction of power and knowledge in the natural sciences. This goal certainly does not mean that science should be rejected, abandoned, or even necessarily modified in significant ways (if it were even possible, let alone desirable, to abandon science in a society in which it is so intertwined with other practices). The positive and productive aspects of power cannot be neglected and should not be undervalued. The aim of this criticism is primarily to *situate* the sciences within the configuration of practices that shape our shared possibilities today, in order to provide a better understanding of our situation—a project that can barely be begun in this book. In the final chapter and an epilogue, however, I shall make such a beginning toward a critical and political understanding of science as a form of power and sketch out the questions that must be asked, however little I have to offer in the way of answers.

Chapter 7

Science and Power

How might the preceding analysis help us reinterpret the political significance of the social practices that constitute the modern natural sciences? At the beginning of the book I made some suggestive but largely undeveloped remarks about power and its interaction with scientific knowledge. To do justice to the political effects of the sciences, I said, we need an alternative to traditional liberal and Marxist accounts of power. Such an account would emphasize the productive character of power rather than repression and distortion; it would focus upon its engagement with actions and practices rather than beliefs; and it would describe its local, decentralized, and nonsubjective deployment. Above all, power must not be characterized as antithetical to or separable from knowledge. The development of knowledge may introduce new forms of constraint, while the imposition of power may itself produce knowledge.

In an important sense, however, this account of power must supplement the traditional accounts rather than replace them, because the traditional theorists of power concerned themselves with themes quite different from mine. They aim to understand the power possessed and exercised by individuals and social classes, and by institutions such as the state, political parties, and the institutional church. Such power is exercised discontinuously on specific occasions and directed toward specific individuals. Typical examples of its deployment include the enactment and implementation of laws, the prescription and execution of punishment, and the formulation and administration of policy. Struggles for this kind of power take place in such explicitly political activities as campaigning and voting, lobbying, arguing in court, and writing and enforcing regulations. Let us, following Foucault, call this explicitly political domain the exercise of "juridical power."[1]

In eschewing primary concern for the place of science in the system of juridical power, I am bypassing what is usually regarded as the politics *of* science. I will not be asking how much power or what kind of power scientists or scientific institutions possess or exercise juridically, and I will certainly not be claiming that scientists possess "too much power" in contemporary society or not enough. These questions concern the political influence of scientists when they stop doing science per se and engage as scientists in such activities as lobbying, testifying, and formulating and administering policy. I want to talk instead about the political effects of what scientists do when they *do science* rather than acting like politicians.

I am equally unconcerned with the effects of traditional political activities upon the organization and practice of scientific research. Government and quasi-government organizations undoubtedly have a major impact upon the practice of science. They support it financially and administratively, deploy scientific resources to serve particular ends of their own (e.g., military or medical), and may proscribe or regulate the practice or dissemination of certain kinds of research. These various interactions between science and juridical power are important and interesting, to be sure, but a focus upon them may mask different kinds of power relations that traverse the very practices of science. Eventually, the interaction between the forms of power I am trying to disclose in this book and the more traditional politics of science needs to be clarified. We need to know how they reinforce one another and what happens when they conflict. We obviously cannot do this, however, until we learn to recognize the kinds of power relations in science that escape the conceptual nets of our most prominent political theories. Achieving this recognition is the aim of my argument in this chapter.

What, then, do I mean by "power," if not juridical power? It will help to recall the discussion of Heideggerian practical hermeneutics at the end of chapter 3. There I argued that all interpretation (which includes all intentional behavior, not just discourse) presupposes a configuration or field of practices, equipment, social roles, and purposes that sustains the intelligibility both of our interpretive possibilities and of the various other things that show up within that field.

1. Foucault 1980: 88.

I will be characterizing power as a characteristic of this field or configuration rather than as a thing or relation within it. Power has to do with the ways interpretations within the field reshape the field itself and thus reshape and constrain agents and their possible actions. Thus, to say that a practice involves power relations, has effects of power, or deploys power is to say that in a significant way it shapes and constrains the field of possible actions of persons within some specific social context. In the previous chapter I used Charles Taylor's account of the importance of such field effects to argue that we must understand science politically. I shall now argue that these political effects can be usefully understood in terms of power. This approach considerably broadens what counts as "political," but it reflects a growing realization by social and political theorists that the everyday practices of seemingly "nonpolitical" agents and institutions are politically significant. To overlook this and equate the domain of the "political" with the realm of the state is to ignore crucial features of the organization and administration of modern societies.

My attempt to include the natural sciences among these more broadly political practices relies extensively upon my argument in chapter 4. There we saw how the development of scientific knowledge is rooted in the construction and manipulation of phenomena through which we develop new skills and uncover new truths and possibilities for truth. We also saw how these developments become disseminated into the world outside the laboratory by standardizing scientific techniques and equipment and by adjusting nonscientific practices and situations to make them amenable to the employment of scientific materials and practices. The result is that the world is increasingly a made world, in the sense that it reflects the systematic extension of such technical capacities, the equipment they employ, and the phenomena they make manifest.

These arguments were not to be understood primarily as an emphasis upon experimental manipulation and control in science rather than upon theoretical explanation. Even scientific theory turned out to be closely tied to our practical grasp of specific model situations, which we extend to other situations by analogy. Science provides us with a transformed hold upon the world, both materially and conceptually. This new grasp in turn transforms the world. For as we saw in chapter 5, what there "is" cannot be intelligibly separated from what we can encounter through the successes and failures of specific practical engagements, where scientific theorizing itself is among these practices.

What we need to understand now is why these earlier claims would be illuminated by interpreting them in terms of networks of power relations. My argument will begin by noting extensive parallels between the construction and manipulation of laboratory microworlds and the various "power/knowledge" relations that Foucault claimed have been at work in numerous modern "disciplinary institutions": prisons, schools, hospitals, armies, factories, and so on.[2] Pointing out these parallels will take up the first two sections of the chapter. By themselves, these parallels would be insufficient to justify describing the natural sciences as traversed by power relations, since it may make a crucial difference that these other institutions aim to reconstruct and normalize human beings, whereas laboratory practices engage very different sorts of things. Foucault himself has claimed that we must respect a distinction between powers deployed against human bodies and the capacities we exercise over things.[3] Of course, the discussion in chapter 6 may encourage skepticism about the import of this distinction, but the point needs to be argued specifically here.

To develop this argument, we must turn in the third section to a further study of the ways scientific practices, equipment, and capacities are extended "outside" the laboratory. Here we can see concretely how scientific practices and the knowledge they embody can act as forms of domination and constraint, helping *produce* us as the kinds of persons we are. We can then see these forms of power already operative within the laboratory itself. In the final section, we can return to the question why the concept of "power" is so important. I shall argue that the tactics for the construction, manipulation, and control of phenomena within the laboratory must be seen as part of a network of power relations running throughout modern societies. The activities in the laboratory that make possible scientific knowledge also directly involve forms of constraint upon people. Further constraints must be imposed in order to extend that knowledge outside the laboratory. They interact with and contribute to the tactics and relations that Foucault has shown to emerge from the disciplinary institutions. They can also be seen to have important power effects of their own. These arguments will show that the laboratory must be understood as another institutional "block" within which and out of which power relations shape us as subjects/agents. Laboratory science and the theoretical reflection it engenders draw upon and reinforce these forms of power in ways that are politically significant.

2. See especially Foucault 1973, 1975, 1977, 1978, 1980.
3. Foucault 1983: 217–19.

Foucault and the Modern Forms of Power

First, we must sketch out the relevant features of Foucault's account of the new shapes of power in the modern world. Foucault has given us not a general theory about power and how it operates, but some perspicuous examples of power relations at work, with certain characteristic features they display. Foucault takes all power relations to affect the body, but he thinks the characteristically modern forms of power target the body on a different scale than previously. "It was a question not of treating the body, *en masse,* 'wholesale', as if it were an indissociable unity, but of working it 'retail', individually; of exercising upon it a subtle coercion, of obtaining holds upon it at the level of the mechanism itself—movements, gestures, attitudes, rapidity: an infinitesimal power over the active body."[4] Instead of being locked away, hidden, the body was made visible and carefully scrutinized; instead of being tortured, it was programmed and exercised; instead of its simply being placed in servitude, its activities were reconstructed for efficiency and productivity. The result was a "political anatomy of detail"[5] through which the body became a carefully crafted instrument and an object of new forms of knowledge. As we shall see, Foucault finds in each detailed tactic of constraint and utilization a new knowledge of the body as well as a new exercise of power.

The gradual refinement of the scale at which power impinges on the body can be seen in several forms. Consider the temporal structure of some of the constraints imposed upon us. Earlier, one might simply have been set a task and left to do it. Then one's performance of the task might have been scheduled: timetables allot a specific time to each specific activity, without deviation or distraction. Finally, the activity itself might be broken down and programmed into a sequence of distinct steps to be performed in order. Now the aim is not just to eliminate the "waste" of time, but to employ productive time more effectively. These programs themselves can then be refined to specify the activity in still greater detail. In this way, Foucault tells us, "a sort of anatomo-chronological schema of behavior is defined. The act is broken down into its elements; the position of the body, limbs, articulations is defined; to each movement are assigned a direction, an aptitude, a duration; their order of succession is prescribed. Time penetrates the body and with it all the meticulous controls of power."[6]

4. Foucault 1977: 136–37.
5. Foucault 1977: 139.
6. Foucault 1977: 152.

The same progression in the detailed exercise of power can be seen in the genesis of skills. Apprenticeship to a master once had a specified duration but a variable course of activities. Apprentices did whatever they were told for so many years, during which time they were expected to acquire and refine the requisite skills of a craft. This variability was replaced by the differentiation of distinct stages of learning: one must master the first stage before learning the second. The staging of the skills could then be supplemented by a program of exercises through which each stage is mastered. One begins exercises not by imitating a specific activity, but by practicing a "component" part and then gradually assembling the components into a single fluid movement, which could then be grafted onto other movements learned in similar ways. In this way the skilled body was literally constructed.[7] As Foucault points out, when this sort of technique has been developed, "one is as far as possible from those forms of subjection that demanded of the body only signs or products, forms of expression or the result of labor. The regulation imposed by power is at the same time the law of construction of the operation."[8] The development of such regulation enables us to understand specific skills and activities in much more rigorous and detailed ways. Its enforcement likewise brings about new knowledge of their practitioners: they are classified in their stage of acquisition of skill and can be differentiated according to their diligence, efficiency, and obedience.

Throughout *Discipline and Punish* and *The History of Sexuality,* Foucault discusses a variety of the strategies and tactics through which such "disciplinary" power is exercised. Here I shall sketch out several strategies that may help us illuminate the power relations that traverse natural scientific research and development. Consider first the tactics of surveillance. Foucault begins by noting an important change in the relations of visibility that mark the exercise of power. In the rituals of sovereignty, power displays *itself* in its majesty and invincibility. Its visibility is spectacular. This display is true of both the triumphal procession of the monarch and the public torture and execution of those who oppose him. The spectacle, however, eventually gives way to practices in which the exercise of power is hidden, while those upon whom it works are increasingly laid open to scrutiny. Visibility becomes a mode of exercise of power rather than just its triumphal display.[9]

7. Foucault 1977: 156–62.
8. Foucault 1977: 153.
9. Foucault 1977: 47–54, 170–77, 195–209, 216–17.

The tactics of surveillance, then, are developed to make the presence and behavior of its objects more readily and completely visible and more accountable in their visibility. Foucault describes an entire architecture of surveillance. His resurrection of Bentham's Panopticon as a model for the ideal prison (or school, workshop, etc.) is the best known in this respect, but he finds this strategy even earlier in the design of military camps in the eighteenth century.

> In the perfect camp, all power would be exercised solely through exact observation; each gaze would form a part of the overall functioning of power. The old, traditional square plan was considerably refined in innumerable new projects. The geometry of the paths, the number and distribution of the tents, the orientation of their entrances, the disposition of files and ranks were exactly defined; the network of gazes that supervised one another was laid down. . . . The camp is the diagram of a power that acts by means of general visibility.[10]

The space within which activity would take place was thus constructed so that nothing significant could be hidden from the gaze through which power was exercised.

Surveillance was not only a matter of architecture, however. The ritual of the examination provided a form of display through which persons could be observed, interrogated, and judged. In the scholastic examination, the pupil's attainment of knowledge itself became the object of knowledge. The physical examination and the taking of a medical history were likewise developed in precise forms to elicit the information needed to judge each individual case. Examinations were also administered in the form of job applications, psychiatric interviews, military reviews, and many other types of inquiry. The display itself produced new forms of appearance that were constructed precisely to render visible some attainment or other characteristic of its object that could not have been readily observed or taken account of without it. These processes enabled people to be classified, assigned, and utilized more effectively, but they also produced new properties for people to possess (a sixth-grade reading level, a history of heart disease, and the like), and with them a knowledge of these properties, their signs, and who possessed them.[11]

Detailed surveillance would not have the same effects at all, however, unless it was accompanied by procedures for tracking, recording, filing, and retaining access to the results of prior observation and

10. Foucault 1977: 171.
11. Foucault 1977: 184–92.

examination. The techniques of writing are closely intertwined with the new forms of surveillance, for the latter are preserved and retain their effectiveness through the archives. "The examination that places individuals in a field of surveillance also situates them in a network of writing; it engages them in a whole mass of documents that capture and fix them. The procedures of examination were accompanied at the same time by a system of intense registration and of documentary accumulation."[12] Foucault claims that both the individual (the "case") and the population, as a differentiated group expressing tendencies and containing distributions, are constituted as objects of knowledge by the various techniques of compiling and manipulating archives.

> Thanks to the whole apparatus of writing that accompanied it, the examination opened up two correlative possibilities: firstly, the constitution of the individual as a describable, analysable object, not in order to reduce him to 'specific' features, . . . but in order to maintain him in his individual features, in his particular evolution, in his own aptitudes or abilities, under the gaze of a permanent corpus of knowledge; and secondly, the constitution of a comparative system that made possible the measurement of overall phenomena, the description of groups, the characterization of collective facts, the calculation of the gaps between individuals, their distribution in a given 'population'.[13]

In Foucault's opinion we cannot fully separate the existence of an object of knowledge from the various practices through which we encounter and deal with it, perhaps especially from the forms of constraint through which it is simultaneously enabled and compelled to show itself in specific ways.

Surveillance is not the only power relation that is built into our physical surroundings. The various ways people are enclosed, grouped, distributed, separated, and partitioned mark a related spatial organization of power/knowledge. These distinctions constrain our patterns of activity and interaction, and in doing so they shape both our activities and us as agents. Whether people are associated by the geometry of their surroundings according to similar ability, similar socialization, or similar function, whether cohesive groups are composed and isolated from others or transient mixtures are allowed continually to form and dissolve, whether people and places are assigned to one another, these spatial distributions shape who we are and

12. Foucault 1977: 189.
13. Foucault 1977: 190.

what kinds of social interactions can take place among us. Spatial distributions function as power relations both physically and conceptually (by constructing the classifications and hierarchies they distribute). Foucault concludes:

> In organizing 'cells', 'places' and 'ranks', the disciplines create complex spaces that are at once architectural, functional and hierarchical. It is spaces that provide fixed positions and permit circulation; they carve out individual segments and establish operational links; they mark places and indicate values; they guarantee the obedience of individuals, but also a better economy of time and gesture. They are mixed spaces: real because they govern the disposition of buildings, rooms, furniture, but also ideal, because they are projected over this arrangement of characterizations, assessments, hierarchies.[14]

These "ideal spaces," the classifications we invent to describe ourselves, we then use as the basis for a variety of "real" differences in our dealings with one another. Classifications provide intelligible ways for people to understand themselves (and the things around them) and to act. The imposition of new categories that come to figure in our various configurations of possible action constrains people in significant ways and even produces different kinds of people. Classification is an important form of power.

We have then, an interaction between techniques of spatial isolation and enclosure (controlled interaction), of nominal classification, description, and explanation, and the previously characterized practices of surveillance and documentation. These techniques function together strategically to administer and organize human beings in more or less coherent ways. But for Foucault a crucial feature of such administration is its normalizing function. It does not primarily operate through the opposition between the permitted and the forbidden, and it does not bring all the weight and power of authority to bear upon those who have been observed in transgression. Its function is corrective. It reconstructs the person and her or his behavior by gradual steps and small impositions. The gaps between observed behavior and established norms are made thematic, and corrective procedures are invoked to bring the offender into line. Normalizing judgment does not aim to abolish deviance, for that would also obviate its own existence. Rather, it creates distributions around the norm, which can be classified, ranked, and dealt with differentially. It both continually

14. Foucault 1977: 148.

corrects and reduces deviation and creates it in new forms so that the norms serve a distributive function.

> For the marks that once indicated status, privilege and affiliation were increasingly replaced—or at least supplemented—by a whole range of degrees of normality indicating membership of a homogeneous social body but also playing a part in classification, hierarchization and the distribution of rank. In a sense, the power of normalization imposes homogeneity; but it individualizes by making it possible to measure gaps, to determine levels, to fix specialities and to render the differences useful by fitting them to one another.[15]

Norms and deviations offer a whole complex field of knowledge and constraint.

We have so far discussed power as if it were simply imposed (By whom? From where?) on inert or docile bodies, which are entirely shaped by the forces that impinge upon them and offer no response or resistance "of their own." There is an important element of this view in Foucault's account of the exercise of power (especially running throughout *Discipline and Punish*), but to leave this picture as final would be mistaken on two counts. It ignores the forms of resistance to power that Foucault thinks are never absent and that make power relations a field of conflict rather than a unidirectional imposition.[16] More important, however, it ignores the productive character of power. It is not just that power relations produce knowledge and transform its objects. They also produce their targets as productive. Foucault insists that the production of discourse and signs is an effect of power relations and, more problematically, that the same is true of their "truth effects." We must consider what Foucault means, and why, when he says that the "will to knowledge" must itself be understood as the outcome of relations of power.

Foucault focuses much of the opening volume of *The History of Sexuality* upon confession as a technique of power. The one who confesses is constrained to produce a discourse about otherwise inaccessible "inner truths." This discourse does not itself reveal that truth directly, however. It is presented to an interpreter (priest, psychoanalyst, parent, teacher, etc.), whose interpretation confers the status of truth upon what has been revealed. Foucault's principal aim here is to challenge the notion that power is fundamentally repressive and that we could be liberated from the effects of power if we could only

15. Foucault 1977: 184.
16. Foucault 1980: 78–92, 1983: 219–22.

express freely what it censors. The "repressive hypothesis" suggests that there is a truth hidden within us, a true self longing to emerge, and a new freedom to be achieved in its emergence.[17] Foucault tries to counter that this emergent truth is the effect of an imposition of power and that the repressive hypothesis is one of the masks behind which this power conceals its impositions.

> The obligation to confess is now relayed through so many different points, is so deeply ingrained in us, that we no longer perceive it as the effect of a power that constrains us; on the contrary, it seems to us that truth, lodged in our most secret nature, "demands" only to surface; that if it fails to do so, this is because a constraint holds it in place, the violence of a power weighs it down, and it can finally be articulated only at the price of a kind of liberation.[18]

At first sight, confession as a technique of power seems to be limited to *speaking* subjects, who are capable of articulating the truths they are allegedly being compelled to utter. But this view is mistaken on two counts, I believe. First, Foucault's analysis can be generalized to encompass any forced production of signs, whether or not they can be heard and understood by the one constrained to produce them. Foucault's general point is that one effect of situating something within a field of power relations is that it may be compelled to reveal itself in new ways, to produce new signs of its presence and behavior. This revelation's being an effect of power is not obviously dependent upon the discursive character of the signs produced. Second, Foucault has anticipated this extension in his insistence upon the role of the mediator of the confession. For it is only through this *relation* to an interpreter that the signs produced acquire the status of a revelation about the true nature, state, or character of their producer. And it is the interpreter rather than the confessor who must be able to articulate these signs in a form that has a truth-value.

> If one had to confess, this was not merely because the person to whom one confessed had the power to forgive, console, and direct, but because the work of producing the truth was obliged to pass through this relationship if it was to be scientifically validated. The truth did not reside solely in the subject who, by confessing, would reveal it wholly formed. It was constituted in two stages: present but incomplete, blind to itself, in the one who spoke, it could only reach completion in the one who assimilated and recorded it. It was the latter's function to verify this

17. Foucault 1978.
18. Foucault 1978: 60.

> obscure truth: the revelation of confession had to be coupled with the decipherment of what it said.[19]

Such signs might also be elicited from things that must forever remain blind to their import. It is true that only in the confession of a speaking subject is the constrained production of signs usually taken to be a source of liberation or reconciliation for the one who produces them. But this is hardly the only way signs and discourse play a political role. It may well turn out that the signs we elicit from things may be at least as politically efficacious as the supposed liberation of inner truths.

There is one final technique of power we need to indicate within Foucault's descriptions before we can go on to consider their possible parallels within the practices of the natural sciences. Foucault assigns an important role to a variety of disciplinary "blocks" within which new tactics of power were first developed and coordinated, and from which they "swarmed" outward to invest social relations and practices not confined within these blocks.[20] Schools, prisons, hospitals, barracks, and factories were both the models and the foci for the spread of disciplinary power throughout modern societies. The discovery that schools or hospitals needed to understand and regulate the entire family and community environment in order to educate or treat their students and patients is symptomatic of this development. Foucault suggests that the same techniques of enclosure, surveillance, administration, normalization, and confession that had functioned within these institutions were also available and adaptable to the micro-regulation of everyday life outside their confines. "While, on the one hand, the disciplinary establishments increase, their mechanisms have a certain tendency to become 'de-institutionalized', to emerge from the closed fortresses in which they once functioned and to circulate in a 'free' state; the massive, compact disciplines are broken down into flexible methods of control, which may be transferred and adapted."[21] The question we must now begin to confront more explicitly is whether it is plausible or appropriate to include the laboratory among these disciplinary establishments within which new power relations were created and from which they emerged to effect important political transformations of the world we live in.

Power Relations within the Laboratory

To understand the laboratory as a field of power relations, we must first return to our discussion in chapter 4 of the creation and manipula-

19. Foucault 1978: 66.
20. Foucault 1977: 209–28, 1983: 218–19.
21. Foucault 1977: 211.

tion of microworlds that exhibit phenomena. These microworlds were the constructed systems within which the objects of scientific research could be made to stand out clearly from the background of other objects and events around them, to appear in new forms, and to be manipulated in new ways. The construction and employment of these laboratory microworlds contain analogues to each of the Foucaultian tactics of power described in the preceding section.

There is the obvious comparison at the outset that the natural sciences have also exhibited a general trend toward finer levels of manipulation and description. Much of the focus of biological research has shifted from whole organisms and their behavior to the functioning of organ systems, then to cells and their interior structure, and finally to the biochemical processes taking place within and between cells. Chemistry began with studies of "substances" (or their qualities) and their interactions, then discovered that these interactions could be studied at the atomic and molecular levels, and finally worked with the quantum theoretical processes within and between the fine structure of atoms. Physics has exhibited a similar move from the atomic to the nuclear to the subnuclear level of analysis. Of course these have not always been smooth, unidirectional, or complete transformations, but the general direction over the last two centuries at least has been unmistakable.

How has this increasingly fine structured understanding and manipulation of things taken place? There is no single, simple story to answer this question, but certainly the constructive experimental techniques that enabled these fine structures to manifest themselves and be available for manipulation and control have been crucial. The same is true of the theoretical interpretations that both followed upon and stimulated these constructions. These techniques are the analogues to Foucault's tactics of power.

Consider the role of spatial enclosure and partitioning. I pointed out in chapter 4 the importance of isolating the microworld of an experiment from unaccounted causal interactions. A variety of equipment and technical skills are used to separate the experiment from other things in the laboratory around it. It is this separation that makes the experimental system a describable microworld in the first place. The point is not that the experiment is partitioned from any possible external influence, which would of course be impossible, but that it must be isolated from those interactions that, in the experimenter's practiced judgment, would influence the processes she aimed to study. But this separation is only one of the many spatial divisions introduced into the laboratory. The laboratory is a carefully structured space. There are bench spaces, spaces marked out for cleanup and disposal of materials and equipment, storage space, instrument places, office/desk space, and so on. There are also means for the separation and mixing of materials: the containers in which specific substances are enclosed and isolated from contamination and the vessels in which they are combined to achieve specific desired effects. There are sterile areas, temperature-controlled areas, negative pressure areas, and radiation areas. And these are of course only a small sample of the spatial divisions that run through the laboratory space. If this structuring seems trivial epistemologically, remember that unless these separations are established and enforced with sufficient care, no statement can be justified on the basis of the work done within the laboratory. They are the sine qua non of experimental science.

These partitioned spaces are also structured to make possible the surveillance and tracking of what goes on within them. Of course, the outcome of an experiment has to be made visible in some form. But as much as possible, scientists also want to be able to follow what is going on throughout the process. As I noted in chapter 4, scientific surveillance is not just a matter of seeing and recording the data. Scientists need to get a feel for the equipment, get their "hands on" the procedures, be able to adjust and refine the various intermediate steps, and so forth. The particular objects and materials of research often are important only because their peculiarities are well understood. Creating an appropriate, familiar, and readily usable experimental system is a highly valued achievement in science. We have already seen that this sort of local knowledge is crucial to the development and extension of our scientific capabilities, including the ability to make things accessible to new levels and kinds of description.

Like the multiple surveillances Foucault sees investing our social interactions, surveillance in the laboratory brings with it an expanded field of writing. Once again, data texts are the obvious case, but we must also remember the naming, labeling, filing, tabling, and so forth that enable scientists to keep track of what has been separated or combined, how it has been manipulated, and what they can claim has occurred as a result. Along with this record keeping has come the proliferating *classifications* with which scientists establish and document identities and differences within the materials they work with. This is perhaps the most difficult to understand as a power effect, because of the "retrospective realism"[22] which almost inevitably invades our interpretation of what we do. The categories in terms of

22. I take this phrase from Pickering (1984).

which we deal with things successfully seem in retrospect to have been forced upon us by the things themselves. But this view neglects the role of our actions and judgments in constructing the context within which the identities and differences we employed could be significant to us. As we saw in chapter 5, it is not the "reality" of the properties corresponding to our classifications that is in question here. The point is rather that this reality itself is the outcome of our activities and classifications. To take a different example than we have used before, one need not doubt the existence of the bewildering array of overlapping kinds of particles uncovered by high-energy physics (hadrons, leptons, fermions, bosons, baryons, mesons, etc.) to suggest that their existence is intertwined with the interests and practices of physicists. That the fermions in their abundant variety "differ" from the bosons is unintelligible without reference to the practices and distinctions through which we assign and take account of quantum spin numbers. This connection was the point of my criticism of Hacking's "experimental realism" at the end of chapter 6. Even in the natural sciences, our practices are responsible for the *intelligibility* of the kinds of things there are, including what counts as identities or differences between them.

We do more in the laboratory than just partition, observe, classify, and record things, however. We construct and decompose them and intervene in their doings. There are substances, particles, processes, organisms, reactions, and mutations that are laboratory products. These and other things brought into the laboratory then manifest themselves and interact in a host of "nonnatural" ways. Laboratories are places where things are made to happen or even to exist.

The standardization of problems, materials, procedures, and equipment that figured so prominently in chapter 4 provides an interesting counterpart to the processes of normalization that Foucault found in the disciplinary deployment of power. At first glance these look quite different. Foucault described normalization as primarily a form of judgment. The construction of norms enabled people to be judged for their conformity or deviance and dealt with accordingly. These judgments were in turn connected to corrective forms of punishment that permitted the detailed reconstruction of those subjected to normalization. Normalization simultaneously helped create social homogeneity and enabled differences and deviance to be noted, assessed, and taken account of. Standardization, by contrast, has nothing to do with judgment or punishment. It is a way of adapting one's procedures and materials to more generalized use and less sensitive and fragile performance.

These differences, undoubtedly important, nevertheless conceal overlapping functional roles. Standardization plays a normalizing role in scientific practice. It adjusts the work of different scientists and groups to one another, disposes of some deviant results, and helps shape judgments of reliability. The acceptability of nonstandard techniques must be justified explicitly, and an important part of the justification in the long run will be the development of standardized versions. Scientific practice *introduces* order into the phenomena it describes; it does not just find order there. And like normalization, it not only reduces disorder, it also makes available new fields of disorder to be ordered and classified.

At the same time, Foucaultian normalization plays an important role akin to standardization. It creates functional, productive people who fit reliably into the social interactions characteristic of productive life. People are made more generally useful and less maladapted to the functional demands of modern life.[23] Like standardized tools and procedures, normalized people are productive in a variety of situations other than the ones they came from.

But perhaps the most important power relation embedded within the laboratory is the production of signs. The objects of investigation are subjected to many manipulations whose aim is not to change how they are, but to reveal it. This view is a Foucaultian reading of what Hacking called the "creation of phenomena." Much of modern scientific practice consists not of making the objects of scientific investigation and their behavior directly visible, but of inducing them to produce signs. Think of the multifarious ways scientists have induced mute and hidden things to "speak" to us of themselves: radioactive labeling, bubble chambers, X-ray crystallography, the various forms of chromatography and spectroscopy, the indirect visibility of sophisticated microscopy and telescopy,[24] and so forth. The vast modern outpouring of confessional texts that Foucault takes note of in *The History of Sexuality* is undoubtedly matched by the plethora of signs by which things mark their presence, their structure, their behavior, and their transformations within laboratory microworlds. The proliferation of the scientific journal literature is well known and often lamented. This literature is, however, a distillation and interpretation of the still more overwhelming production of data texts that emerge as a constant flood within the many functioning laboratories of the world.

23. Foucault 1977: 210–12, 218–28.
24. On the question of visibility through microscopes, see Hacking (1983, chap. 11).

This suggested parallel between the confession, in which persons are induced to articulate their innermost thoughts, desires, and feelings, and the inscription devices[25] through which the objects of scientific research produce signs of themselves may still seem farfetched despite the arguments in chapters 3 and 6 that aimed to dispel the classical hermeneutical distinction between the interpretation of meaningful texts and the explanation of how things behave. The introspective dimension of the confession has no counterpart in the constrained production of signs within the laboratory: molecules and particles have neither inner secrets to confess nor a language in which to confess them.

Once again, however, if we examine Foucault's account of confession as a technique of power, this difference becomes largely irrelevant. I argued in the preceding section that the *discursive* character of signs is not necessary for signs to be the outcome of power relations. Furthermore, Foucault would certainly insist that the form of the confessions produced by speaking subjects has less to do with their privileged relation to some inner truth than with the particular power relations within which they are produced. Instead of appealing to an inner content that freely emerges in the confession, one would do better to look to the codification of the sacrament of penance, the texts of psychoanalytic theory, and popular confessional literature if one wishes to understand what would be confessed and what form the confession would take. These kinds of texts exemplified what could *count* as a true confession, in both senses of the word "true." Just as garbled or randomly scattered data plots often indicate an unsuccessful experiment rather than an effect that needs to be accounted for scientifically, the "confession" that does not say something deemed to be revealing does not count as a genuine confession. The hermeneutical dimension of the production of signs is thus to be found primarily in the practices that elicit and interpret them rather than in the person or thing that produces them. This circumstance is true whether these practices involve placing a person on a couch and recording and interpreting her or his reports of dreams or placing an object in front of a beam of accelerated particles in the midst of apparatus to detect the scattering of the beam.

Even if one were inclined to accept all these claims about scientific practice, however, the question would remain why we should regard these parallels between laboratory practices and the tactics of disciplinary power as more than amusing coincidences. After all, power is

25. Latour and Woolgar 1979: 45–69.

a *political* concept, having to do fundamentally with relations between people. We may speak metaphorically of power over nature, but perhaps this makes no more literal sense than to speak of our sovereignty over nature or our government of it. We should not allow ourselves to be so carried away by Foucault's expansion of the political domain that we forget what politics is fundamentally about—namely, the governance of human beings and their institutions.

The response to this concern is to be found in the laboratory analogue to what Foucault called the "swarming" of the disciplines, their emergence from the enclosed institutions within which they were constituted into the surrounding body politic. The laboratory is not an enclosed and isolated ivory tower but an important political force in the modern world. We saw in chapter 4 that science does not merely deliver new ideas, devices, materials, and techniques to a world that is otherwise unchanged. Scientific practices and achievements are a powerful force in remaking the world. On the one hand, particular scientific concepts and discoveries themselves have a significant impact upon the field of possible action within which we find ourselves. Perhaps more important, however, the general efforts required to make our world amenable to a variety of scientific innovations, and to the continuation of the scientific practices themselves, merge with and reinforce the sorts of disciplinary strategies and tactics described by Foucault. The relations between scientists and their materials and objects of study count as power relations not only because of what happens specifically in the laboratory, but because of their political effects in other contexts.

Science as Power outside the Laboratory

Even to begin seriously to inventory the power relations through which scientific practices and achievements impinge upon the possible actions of us all would be an impossible task for the scope of the present inquiry. I propose instead to suggest and exemplify some important general features of the power effects of scientific practices, which I hope will suggest how pervasive and significant they are. Bruno Latour has recently remarked that "in our modern societies, most of the really fresh power comes from sciences—no matter which—and not from the classical political process."[26] My modest aim in this section is to say enough to convince the skeptical reader that Latour's claim is not hyperbolic and to convey some limited sense of the scope, direction, and character of the power relations whose ori-

26. Latour 1983: 168.

gin can be traced back to laboratories. In the following section I will explain *why* it is important to think about these relationships in terms of power.

The most obvious and apparently simple effects of the sciences upon everyday social and political practice stem from the transfer of specific new materials, processes, and devices from the laboratory context into the world "outside." It is not difficult to think of just a few such innovations that have radically transformed what we do, not just opening up new possibilities but compelling others upon us and foreclosing the utility or intelligibility of still others. I would initiate such a list with the creation of sustained electrical currents and their various transformations into work and heat; the synthesis or isolation of hundreds of thousands of new substances through synthetic organic chemistry and petrochemistry; the transmission, reception, and processing of electromagnetic radiation, whose variations in frequency and amplitude convey information; the achievement of sustained nuclear fission chain reactions; and the invention of pharmaceutical, surgical, and other means of medical intervention. The impact of these alone has been immense, and yet the list could be continued almost indefinitely. Even if one gives only the narrowest characterization of the political realm, these developments have significantly transformed the central preoccupations of political life and the means by which we address them. The stimulation, protection, regulation, utilization, and remedying of the effects of these translations of laboratory practices, materials, and procedures now occupy a large proportion of our overtly political activity. If one broadens the "political" realm to include all forms of constraint and governance of the field of possible action, as I shall do, the political impact of these laboratory transfers is all the more overwhelming.

We cannot confine the political impact of laboratory practices to the direct effects of extending particular achievements outside the immediate laboratory context, however. As I noted in chapter 4, one cannot extract particular items or procedures from the laboratory and expect them to function reliably, or even predictably, in a different environment. One must to a certain extent extend the laboratory context itself if this transfer is to be successful. In my earlier discussion I focused upon the example of the nontrivial effort required to provide and apply common measures of length, time, weight, temperature, and the like wherever one wishes to transfer laboratory achievements effectively. This requirement, however, is only one of the many imposing preconditions of such transfer. A few more examples may help indicate the magnitude and character of the effort required.

Many of the isolations and separations of materials that enabled the achievement of effects within the laboratory must be partially extended if those effects are to be reliably sustained elsewhere. Admittedly, the standardization that desensitizes these achievements even for ongoing laboratory use is usually increased significantly in such cases. Nevertheless, purified and measured substances must be protected from contamination, machines must be kept clean, oiled, supplied, and maintained and must be operated "properly," their environment must be controlled for temperature, humidity, sterility, and so forth. These separations must even be maintained to some extent after use; think of the effort required to prevent newly created toxic substances from entering groundwater, the atmosphere, or the food chain or to remove them once they get there.

All the preconditions mentioned so far—namely, the maintenance and application of standard measures, the purification, isolation, and availability for use of laboratory materials, and the construction and proper maintenance and use of the requisite machines and instrumentation—require the extension of laboratory-like practices of tracking, recording, labeling, filing, and accessing these things in written documents. One need only think of the effort required to determine and keep track of the chemical constituents of the vast variety of measured substances now available to us, from foods and drugs to cleaning compounds and pesticides. These efforts range in seriousness from the necessarily obsessive (but unsuccessful) attempts to keep track of the location and disposition of every fraction of an ounce of weapons-grade plutonium or uranium-235 produced in this country, to the much more haphazard accounting of materials that predominates among the makers of chemicals, processed foods, and drugs, to the occasional fabrication of figures to preserve at least the appearance of documentation. But there is little doubt that without this massive accounting effort, we would fail to preserve the separations and mixings, the counting and the measurements, that are essential to extending scientific practices and achievements outside the laboratory.

It is important to recognize that I am not trying to claim that these various practices originated in laboratories or that there is a one-directional relation between their employment in laboratories and their application in other contexts. Clearly there is a lively and complex interaction between the practices, techniques, and equipment used in laboratories and those found in other work contexts. Laboratory life has been greatly enriched by imports from other institutions and practices. Also, many of the practices that I have described as characteristic of and essential to the laboratory have a much longer history: writing, keeping accounts, keeping time, purifying and isolating substances, the now industrialized practices of glassblowing and metal fabrication, and so forth, originated earlier in rather different contexts and for a diversity of ends. Even the attempt to construct experimental microworlds can undoubtedly be discerned before there were any facilities recognizable as laboratories in their modern guise. The laboratory, then, represents a mature form, not the origin of the gradually emerging project of remaking the world to make it knowable. The laboratory is a useful focus for our understanding of this project, however. The ensemble of practices, skills, and equipment that come together in laboratories gives these older activities a new sense, comparable as I noted earlier to Foucault's account of the constitution of the ordinary individual as an object of surveillance and description. Lavoisier's attempt to enclose and isolate combustion so that all of its products could be collected and weighed was an unusually clear early manifestation of this new project: these particular things and the processes they underwent were important because they could be minutely and precisely known, rather than becoming known because they were important. The laboratory is a symbol rather than a cause of this reversal in the field of knowledge.

A related caution should be recalled from my discussion in chapter 4 of theorizing and experimentation. When I speak of laboratory science, I am not trying to contrast it with theoretical science, or even with field research or astronomical observation. The creation of phenomena in the laboratory and the attempt to improve the theoretical understanding of the things that are manifested there function together: scientific understanding draws its sustenance from their interaction. And although there are undoubtedly some interesting differences between laboratory work and field research in biology or geology, these differences largely reflect the adaptations needed to achieve similar effects in a different setting rather than a fundamental difference in aims and practices.

So far we have spoken of the summing of many individual preconditions that must be imposed upon us if we are effectively to extend scientific knowledge beyond the constructed microworlds of the laboratory. But very few scientifically developed materials or procedures can be utilized in isolation. In this respect they are like any other equipment. Yet in many cases the products of scientific/technical work function effectively only in concert with other technically generated products. This point is important because the environmental controls needed for them to function reliably may interlock. The constraints on their environment and the behavior required to create and sustain it that are imposed by each individually may be minimal, but the overall effect may be to impose substantial constraints upon human action. Langdon Winner has noted that "certain kinds of regularized service must be rendered to an instrument before it has any utility at all. One must be aware of the patterns of behavior demanded of the individual or of society in order to accommodate the instrument within the life process."[27] The networks of equipment and practice that have their origins in the carefully constructed, controlled, and surveilled microworlds of the laboratory may well impose more extensive and stringent burdens upon us than do other kinds of tools or procedures. There may even be systematic constraints that the remaking of the world to resemble laboratory microworlds may impose upon us. And if so, these would certainly count among the power effects of the laboratory as a disciplinary block.

I shall focus upon two characteristic and interrelated features of these *systemic* constraints that seem to be power effects of the extension of laboratory discipline. The first is an increase in the interactive complexity and "tight coupling" of the technical systems that extend laboratory achievements and of the social organizations that manage them. These terms were introduced by Charles Perrow to illuminate the occurrence of what he calls "normal accidents" in certain high-risk technological systems. These are accidents that are due not so much to the malfunction of a single component of a system as to multiple failures whose combination was not anticipated. He claims that such accidents are to be expected in systems that are complex and tightly coupled.[28]

"Complexity" refers specifically to a variety of nonlinear connections in technical systems and their controlling organization: for example, common-mode components (components that serve multiple functions), physical proximity of different subsystems, indirect sources of information for monitoring the system, and vulnerability to changes in the system's environment.[29] From the perspective of the operators of technical systems, "*complex interactions* are those of unfamiliar sequences, or unplanned and unexpected sequences, and either [are] not visible or [are] not immediately comprehensible."[30] Although Perrow does not make the link specifically, these interactions are frequent and sometimes necessary features of the extension and expansion of laboratory microworlds. Indeed, his most salient examples of interactively

27. Winner 1977: 194–95.
28. Perrow 1984, esp. chap. 3.
29. Perrow 1984: 72–86.
30. Perrow 1984: 78.

complex systems—nuclear power plants, chemical production, and industrial-scale biotechnologies—are clearly laboratory-derived processes. The indirect sources of information that are an important part of the increase in complexity are themselves, of course, the industrial analogues to the constrained production of signs in the laboratory.

"Tight coupling" means a lack of flexibility in the connections between technical and/or organizational subsystems: "There is no slack or buffer or give between two [tightly coupled] items. What happens in one directly affects what happens in the other."[31] Tightly coupled processes tend to be time dependent ("they cannot wait or stand by until attended to"),[32] constructed with invariant and noninterchangeable sequences built into them and with minimal slack in their use of resources ("Quantities must be precise; resources cannot be substituted for one another; wasted supplies may overload the process; failed equipment entails a shutdown because the temporary substitution of other equipment is not possible").[33] Once again, these are characteristics of constructed microworlds; the clear, regular causal connections established in the laboratory owe their clarity and regularity to the careful monitoring and control of the timing, sequencing, and material inputs of the processes studied. There is some flexibility in these matters, and good experimental judgment allows a scientist to know about how much, but generally the importance of good technique in research testifies to the need for precise control of inputs and processes. Standardization routinizes much of this control and to some extent desensitizes the processes, but much of the tight coupling of these processes is intrinsic to their success.

The interactive complexity and tight coupling common to the technological extension of laboratory microworlds are only one aspect of a twofold effect, however. For while our technical constructions become more complex and tightly coupled, the natural environment is artificially simplified, controlled, and stripped of some of its own capacity for self-regulation and buffering. The simplification of the environment, or else the isolation of a constructed microworld from the complexity and incomprehensibility of environmental interaction with it, is crucial to the clarity and regularity of the phenomena produced in the laboratory. Drastic steps may be taken to achieve this simplification in the context of scientific research. Among the examples we discussed in chapter 4, we saw Raymond Davis's massive ex-

31. Perrow 1984: 90.
32. Perrow 1984: 93.
33. Perrow 1984: 94.

perimental apparatus for neutrino detection buried miles underground in a salt mine to screen out unwanted forms of radiation that would have produced too much noise in the detector, and we saw Roger Guillemin and Andrew Schank process millions of animal hypothalami to exclude the "contaminating" substances that provided the normal functioning environment for the releasing hormones they were investigating. Usually one must drastically simplify the environmental background if the structure and mode of action of a substance or the steps of a process and its causal antecedents are to be clearly revealed.

The attempt to extend scientific knowledge and skills outside the laboratory necessitates that this environmental complexity be reengaged to some extent. Nevertheless, this necessity often requires simplifying the natural environment rather than adapting scientific practice to a more complex environment. As an example, consider the "green revolution" in agriculture.[34] This development was the product of intensive agricultural research aiming to increase the yield of staple food crops. Among the techniques used to achieve higher yields were the introduction of hybrid strains of the crops that generated the highest yields under experimental conditions; the intensive use of carefully calculated amounts of chemical fertilizers; the employment of irrigation systems to regulate precisely the amount of water the crops received; and the intensive use of pesticides to control infestation.

These techniques were not independent. Compared with locally occurring varieties, the hybrid strains consumed more nutrients, were meant to be farmed more intensively (preventing natural regeneration of the soil), and were more sensitive to the soil's chemical balance, thus requiring fertilization. They similarly required a more carefully controlled water supply; and as a monoculture, without the genetic diversity to be found in natural crop strains, they were more vulnerable to blight and infestation. Thus, as several critics have noted, the new plant hybrids were not really high-yield varieties, but high-*response* varieties: changes in their environment and the inputs to it had a greater effect upon them. The high yields produced in experimental studies could be duplicated in the field, but at the cost of removing natural barriers to failure such as genetic diversity and hardier stocks and of requiring energy-intensive and capital-intensive inputs to create and maintain the necessary environmental conditions.

34. My discussion of this example draws extensively from Frankel (1971), Lappe and Collins (1977, esp. part 5), Scott (1976, esp. chap. 7), and Commoner (1976: 149–64). I am grateful to Catherine Newbury for bibliographical advice in this area.

The result was that agricultural practice was *forced* to become more tightly coupled and artificially complex, since the natural complexity that had allowed the system to be more loosely coupled had been destroyed in order to apply the knowledge developed in experimental studies. It is crucial to emphasize the connection between power and knowledge here. The very constraints that enabled us to keep track of what happened in the field and to know the relations between inputs, environmental conditions, and crop yields were those responsible for the tight coupling and the artificial complexity/natural simplicity of the new practices. Nor is this relationship accidental. Tighter coupling and the *control* of what complexity there is are necessary conditions for a detailed knowledge of processes and causal connections.[35]

A similar cycle, in which the attempt to gain more knowledge of and control over natural processes removes the slack that was originally there and thereby necessitates more tightly coupled technical intervention, is characteristic of many medical specialties. Obstetrical practice provides a good example. Hospital birth provides a more carefully controlled environment for monitoring the progress of labor and the condition of the fetus. Electronic fetal monitoring keeps track of the baby's heartbeat. Intravenous feeding controls the mother's caloric intake and makes possible general anesthesia for surgical intervention (as does the hospital environment generally). Other anesthesia is available to control the pain of labor. Oxytocin or other drugs can be used to increase the pace of contractions. If there are signs of fetal distress or insufficient progress in labor, mechanical extraction or cesarean section may be performed. Intensive care facilities are then available for premature, injured, or diseased newborns.

But once again, these techniques and procedures are not independent of one another. Efficient use of facilities and personnel encourages the use of oxytocin and surgical intervention to control the timing of labor and birth. Both of these practices increase the need for anesthesia. Fetal monitoring can raise the likelihood of the fetal distress it is supposed to recognize, both by lessening maternal mobility, thereby decreasing blood pressure and the oxygen supply to the fetus, and by increasing the risk of umbilical cord prolapse. Fetal monitoring is also associated with a significant rise in the performance of cesarean sections. All these techniques (including hospital delivery itself) intensify the risk of infection and the need for antibiotics to

35. This is true even if, as Perrow 1984 plausibly argues, tight coupling and complex interaction together also make for uncontrolled interactions and new, unknown causal connections; see references in note 34 above.

suppress it. Most of them also increase the frequency of premature birth and the need for intensive neonatal care. The very procedures that enable us to monitor and control the birth process thus also make it more dependent upon the use of those procedures.[36]

The examples given so far might once again suggest that the power effects of scientific knowledge result primarily from its technological application. But this idea would be mistaken on two counts. First, the implicit separation between the achievement of knowledge and its subsequent application has already been shown to misunderstand scientific practice. Scientific knowledge is not something distinct from our practical ability to manipulate and control the phenomena it interprets. A local and practical grasp of specific situations and phenomena is an important part of scientific knowledge. Generalizing that local knowledge is more a matter of extending specific practical skills and reconstructing or reconceptualizing new situations to make them amenable to the extension than of discovering and applying universal laws. Even theoretical knowledge is tied to its treatment of specific model situations, which can be extended to other situations by analogy. The result is that if the technological extension of scientific knowledge must be understood in terms of power, so must the initial acquisition of that knowledge.

The second problem with restricting our understanding of scientific power to its technological application is that power effects arise from science that cannot be traced to any particular technological extension of it. Scientific knowledge itself changes our practical understanding of the situations within which we act. Consider first an example continuous with the cases we have just been discussing. A more detailed understanding of nutritional biochemistry has changed the way we think about diet, regardless of any technical innovations in how we grow and prepare foods. We now think about nutrition compositionally—that is, in terms of carbohydrate, protein, fat, fiber, vitamin, and mineral composition—and this tends to change what and how we eat. Cognitive science may be having similar effects upon our thinking. The attempt to describe thought in terms of decision procedures does not merely mirror processes already going on; it encourages us to *revise* our thinking to match the procedural model. The heuristic strategies, decision trees, and learning programs constructed by cognitive scientists to describe how we think later show up in the educational materials through which various skills and disci-

36. Brackbill, Rice, and Young 1984, esp. chap. 1.

plines are taught.[37] Even evolutionary theory, seemingly the quintessentially not "applicable" theory, rearranges our understanding of ourselves and our situation as selection models, and the notion of evolutionarily stable strategies, change the way we think about political life or epistemology, not just describing our political and scientific practices but changing them.

Nor must it be thought that such effects occur only at the borders of the human sciences. Thermodynamics provides a good example of a physical theory that rearranges the way we deal with the world. Agriculture and architecture are fields of social practice that are vitally affected by thinking of them as systems for using and converting energy. The effect of thermodynamic thinking on architecture should be obvious to us today, but thinking of agriculture as solar energy conversion, and examining its first- and second-law efficiency, provides a powerful critical tool for reconstructing agricultural practices.[38] The moral of these examples is that reorganizing the terms in which we conceptualize a field of practice, as the sciences often enable or compel us to do, can transform that field just as effectively as introducing a new material, device, or procedure.

But these conceptual changes are not limited to the effects of particular theoretical models. The compositional and calculative practices that run throughout many of the sciences exercise more global power effects of their own. The understanding of the earth as a collection of resources, to be compiled, extracted, conserved or exploited, substituted for one another, traded off against one another, and generally used to maximize their net benefit, provides a good example. This understanding has a profound effect upon how we deal with the things around us, and it changes the issues at stake in our dealings with them. Our concerns with efficient utilization of resources, with questions of conservation and exploitation, with trade-offs and substitutions between resources, all have their origins in the laboratory-like understanding of things as isolable and usable under controlled conditions to produce determinate effects.

Analyzing the world into its resource components has a particularly interesting effect upon our self-understanding. We occupy an ambiguous place in such calculations, since we (or at least some of us) presumably are the ones who are supposed to benefit from the prudent use of the earth's resources, but we also count in the calculations as

37. Dreyfus 1979; Dreyfus and Dreyfus 1986.
38. See, for example, Commoner 1976: 149–64.

various types of "human resources" (skills, labor, consumption, genetic diversity, etc.). The ambiguity is particularly evident when what is being calculated is the "carrying capacity" of the planet, or its optimum human population. At this point the supply of human beings is clearly supposed to be what is adjusted as well as who the adjustment is for.

The understanding of actions and events as having calculable outcomes transforms the field of our practices in similarly profound ways. This issue differs from the classical concerns about causality and determinism. The question is not whether events *have* determinate causes, but whether we can discover, disentangle, and influence their causes. Most of the examples I discussed previously in this chapter have been variations on this general theme, reflecting attempts to *make* actions and events calculable. Making the world more predictable, by reducing the chaotic complexity of natural events and processes to regular procedures that can be controlled and taken account of, is an issue in virtually all our scientific practices. This simplification is undoubtedly what Heidegger had in mind when he said:

> Modern science's way of representing pursues and entraps nature as a calculable coherence of forces. Modern physics is not experimental physics because it applies apparatus to the questioning of nature. Rather the reverse is true. Because physics, indeed already as pure theory, sets nature up to exhibit itself as a coherence of forces calculable in advance, it therefore orders its experiments precisely for the purpose of asking whether and how nature reports itself when set up in this way.[39]

The laboratory provides the setting for most scientific work, because the microworlds constructed within it are more amenable to such calculative control. They reveal "how nature reports itself" under these constraints. In turn, they provide both models and concrete capabilities for extending that calculative control beyond the boundaries of the laboratory. Foucault saw the panoptical prison as the implicit model of a "carceral society." The laboratory provides a comparable but more adequately realized model for a calculative world.

Why "Power"?

I have been describing as power relations a variety of effects of the development and extension of laboratory practices. We are now pre-

39. Heidegger 1954: 29, 1977: 21.

pared to consider whether and why the manipulation and control of things in the laboratory (and its associated theoretical understanding) should be understood in these terms. The acquisition of scientific knowledge does not leave unchanged the things known. But it also changes and constrains the people who deal with those things, in ways that cannot be clearly or neatly separated from the constraints imposed upon the things themselves. As soon as laboratories cease to be fairly isolated institutions whose practices and equipment remain enclosed within their boundaries and social networks, the associated forms of constraint also move outside.

Let us begin within the laboratory itself. We have seen the laboratory as a space of stringently enforced enclosures and separations, of strict surveillance and tracking, of carefully controlled interventions and manipulations. These constraints upon the materials and processes that occur within its constructed microworlds cannot be sustained unless there are also constraints (largely self-imposed and self-monitored) upon the persons who work within them. Laboratory practice imposes a detailed discipline upon those who engage in it. This discipline is not normally noticed, because it becomes routine and ingrained in scientists and technicians, who have long since internalized it. It can therefore perhaps best be illustrated at the margins, when it does not work effectively. Consider the following reconstruction of a sociologist's efforts as a laboratory technician:

> One of the most difficult tasks was the dilution and addition of doses to the beakers. He had to remember in which beaker he had to put the doses, and made a note, for example, that he had to put dose 4 in beaker 12. But he found that he had forgotten to make a note of the time interval. With pipette half lifted, he found himself wondering whether he had *already* put dose 4 in beaker 12. He blushed, trying to remember whether he had made a note before or after the actual action took place; obviously, he had not made a note of when he had made a note! He panicked and pushed the piston of the pasteur pipette into beaker 12. But maybe he had now put *twice* the dose into the beaker. If so, the reading would be wrong. He crossed out the figure. The observer's lack of training meant that he continued in this fashion. Not surprisingly, the resulting points exhibited wide scatter.[40]

As this example illustrates, if one is to isolate, track, and record controlled phenomena in the laboratory, one must keep comparable control over the actions of the people who work there. The laboratory

40. Latour and Woolgar 1979: 245.

requires that one be accountable for the least actions that might affect the phenomena being constructed and observed there.

Comparable constraints must be maintained if the effects achieved within laboratory microworlds are to be extended beyond their boundaries. As Latour noted, for example, Pasteur's triumphant demonstration of his anthrax vaccine in a public "experiment" at Pouilly le Fort was preceded by extensive negotiation concerning what would need to be done to enable the trial to work (and to *count* as a genuine success). For without making animal husbandry a less dirty, haphazard, and locally variant activity than it had been, without introducing and enforcing such laboratory practices as "disinfection, cleanliness, inoculation gesture, timing and recording"[41] upon the French farm, the vaccine would not have had demonstrable effect.[42] Pasteur did not merely introduce a new material into an unchanged social setting, he required serious changes in the practices and practitioners surrounding its introduction.

When the equipment and procedures that affect our lives become more complex and tightly coupled, the social relations through which we deal with them must become more tightly coupled as well. In technical systems that are highly sensitive to environmental change or the malfunction of their components, the social organizations that manage them must be responsive to such difficulties and must respond according to the invariant time frames and action sequences built into the systems themselves.[43] If there is less slack, and less margin for error within the technical system itself, then the same will be true of its controlling organization. We must keep more careful track of what the people involved are doing and tie their actions more closely to the demands of the system they are working with. They will undoubtedly be given more information about what is going on around them, but the increase in their knowledge itself exercises a constraining effect upon them, for they are compelled to keep track of that information and tailor their actions in response. If a technical system must be closely monitored to function properly (and even to avoid catastrophe), then those who monitor it must be more tightly constrained as well. We are less free to leave things alone.

41. Latour 1983: 152.
42. Latour 1983: 152.
43. Of course, when *complex* systems are tightly coupled one is occasionally faced with unanticipated and misunderstood difficulties, which nevertheless require prompt, carefully orchestrated, virtually preplanned responses; the organizational requirements of such systems may be internally inconsistent and thus impossible to respond to appropriately. See Perrow 1984, esp. pp. 329–52.

But the increased demand for effective monitoring and control of our environment and of those who monitor and control it is hardly the only power effect of the extension of laboratory practices and materials outside the institutions in which they originated. For it is not only a controlled material environment that is a precondition for the effectiveness of these extensions. The social environment must be reorganized and controlled as well. To see this necessity, let us return to our earlier examples of maximized-yield agriculture and high-tech obstetrical care.

The green revolution's combination of monocultures using new hybrid seeds, intensive fertilization, controlled irrigation, mechanization, and extensive application of pesticides is not applicable within just any social system. These techniques cannot be effectively applied to small acreages. This limitation is partly due to their capital requirements, which are beyond the reach of small farmers, and it also results from their increased riskiness: high-yield techniques run greater risks of a single disastrous year, which larger operations can afford but which are devastating to farmers operating near the margin of subsistence. But even if smaller farmers were subsidized or formed cooperatives to pool capital and absorb risk, the machines, irrigation systems, and pesticides need large fields to be effectively employed. Farmers must also invest substantial time and skill to acquire the understanding and information necessary to use them effectively, an investment that is likely to pay off only on large plots of land (in addition, timely access to the information may require the political connections that frequently accompany greater economic resources). And perhaps most important, these techniques drastically reduce the need for (and therefore the economic value of) human agricultural labor. Their use drives a significant proportion of the rural population from the land, whether they were independent small farmers or agricultural laborers. Even if these techniques could be used effectively on small farms, their use would make the efforts of many of the farmers redundant. The result is a massive reshaping of the social roles and practices of agricultural communities.[44]

Comparably significant effects upon the social organization of an activity result from our increased knowledge of and control over the birth process. The physical setting of birth changes as the hospital delivery room replaces the home. Birth is isolated from the rest of one's life and removed to a setting specifically designated for birthing. This isolation in turn changes the social role of attending profes-

44. See esp. Lappe and Collins 1977, part 5; Scott 1976, chap. 7; Frankel 1971.

sionals; their identity frequently changes also, as midwives are replaced by obstetricians. The role of the mother herself changes, as she is defined as a less active participant in the process. Even when she is not medicated and is fully conscious, her understanding of her situation is subordinated to what the monitoring instruments say is happening. Even her physical posture is changed as the supine position, which is less efficient and comfortable for her, is used to make observation and intervention easier for attending hospital personnel.[45] Once the baby is born, its isolation in the nursery enhances the professional monitoring of the condition of both mother and baby, but their separation fundamentally changes their potential relationship. As a result it is now possible to acquire a much more exact and detailed knowledge of what is going on during and immediately after birth, but the routine use of the procedures required to obtain this knowledge has changed the birth process significantly.

I have not specifically discussed in these examples the increased power and influence accorded to various professionals by such extensions of scientific knowledge. Specialized knowledge and the institutions through which it is deployed often give scientists, engineers, and physicians considerable power over their clients and significant advantages in competition for social resources. These effects are real and important, but they are more widely discussed and better understood than the kinds of changes I have focused upon, because they can be more easily accommodated to a traditional understanding of power. What I have tried to illustrate here is that the context of these more traditionally recognized political changes has already been more subtly altered by the development and extension of our scientific practices. The things around us are more carefully measured and counted, decomposed and reconstructed, isolated from uncontrolled interaction with their environment, tightly coupled and artificially complex. We, our institutions, and our understood field of possible action have been transformed accordingly.

Laboratory practices and their extension also coincide with and reinforce the disciplinary practices and institutions that Foucault has taken pains to characterize. First, the functioning of the laboratory itself, and the extension of its practices and materials outside its walls, would be unthinkable without the imposition (largely self-enforced) of the normalizing disciplines and constraints Foucault was describing. From the educational practices that prepare scientists for their work to the forms of surveillance, normalization, and constraint that

45. Brackbill, Rice, and Young 1984: 12–13.

sustain their work itself and its reliability, the disciplines are crucial to the sciences.[46] We have also seen that they are essential to the various extensions of scientific practice outside the laboratory, which impose upon the world many of the controls and constraints that originated in the laboratory. Second, the microworlds produced, manipulated, and studied in the laboratory are themselves models for a tightly coupled, "microscopically" observed and controlled world.[47] Laboratory experiments come closer than perhaps anything else we know to a world that includes nothing that is not supposed to be there, where nothing happens that is not intended or accounted for and everything is subject to careful surveillance, classification, and documentation. This "ideal" of a completely monitored and controlled world is of course never actually fulfilled even in an experimental microworld, but it is much more realistically approximated there than elsewhere, and it can be seen as a different expression of Foucault's account of a "disciplinary society."

There is a third convergence between laboratory practices and Foucaultian disciplines that is much more important, however. For Foucault, what is most fundamentally characteristic of the disciplinary forms of power that first emerged clearly in the mid-eighteenth century is their strategic aim to increase social productivity and utility. He noted:

> The development of the disciplines marks the appearance of elementary techniques belonging to a quite different economy: mechanisms of power which, instead of proceeding by deduction, are integrated into the productive efficiency of the apparatuses from within, into the growth of this efficiency and into the use of what it produces. . . . These are the techniques that make it possible to adjust the multiplicity of men and the multiplication of the apparatuses of production (and this means not only 'production' in the strict sense, but also the production of knowledge and skills in the school, the production of health in the hospitals, the production of destructive force in the army).[48]

Foucault was primarily interested in the practices through which human bodies were made more productive and useful. But he recognized that these efforts could not be separated from those that aimed to organize and utilize the world around us more efficiently and

46. For a discussion in different terms of the normalization of scientific work itself, see pp. 119–25 above.

47. It is perhaps no accident that the microscope provides both a common symbol for the laboratory and a common metaphor for any form of detailed surveillance.

48. Foucault 1977: 219.

effectively. He insists that the development of the disciplines and the emergence of a capitalist mode of production were closely intertwined, where the relevant defining feature of capitalism for him was not the domination of one class by another, but the accumulation, organization, and productive utilization of capital.[49] One could not reconstruct and organize the population more productively without also reconstructing and reorganizing the material forces of production. Laboratories and scientific practices have played an increasing role in that reconstruction.

It is important to understand that the characteristics and effects of scientific practices I have been describing are not to be understood as the result of some grand conspiratorial design to reshape and normalize our world. They represent a coincidence of overlapping and selectively reinforcing practices and techniques that have their own local rationales for adoption. One can say of them what Foucault said of the political anatomy of the body he saw emerging in the eighteenth and nineteenth centuries: "The 'invention' of this new political anatomy must not be seen as a sudden discovery. It is rather a multiplicity of often minor processes, of different origin and scattered location, which overlap, repeat, or imitate one another, support one another, distinguish themselves from one another according to their domain of application, converge and gradually produce the blueprint of a general method."[50] But their mutual reinforcement plays an important role in their development. Many of the material and conceptual extensions of the laboratory are interdependent; one scientific field may provide the signs that allow another field's innovation to be monitored, the control devices that regulate it, the specialized materials that permit its construction on a new scale or to a new end, or the conceptual realignment that suggested it in the first place.

More important, the laboratory-based practices of measuring, counting, and accounting, of analysis, decomposition, and recomposition, and of controlled separation, isolation, and mixture, in addition to reinforcing one another, provide the basis for understanding the world itself as a collection of resources. The microworlds produced in laboratories similarly provide the material basis for regarding events and situations as analyzable causal composites.[51] All these tactics and

49. Foucault 1977: 220–21.
50. Foucault 1977: 138.
51. Strictly speaking, of course, celestial and terrestrial mechanics, which did not require much laboratory activity, provided the classical model for this approach, but it is the extension of this kind of analysis of causal components to the "Baconian sciences" of chemistry, anatomy, heat, electricity and magnetism, and so on, that was most crucial to the productive utilization of the earth and, concomitantly, its population; this extension was achieved in the specialized settings of the laboratory and the clinic.

interpretations in turn coalesce with the Foucaultian disciplines. The earth as a natural resource and population as a human resource, to be organized and utilized efficiently as productive forces, are correlative notions. The understanding of causality as analyzable is also crucial to the creation of new productive forces, and thus to the interplay Foucault described between the accumulation of capital and the "accumulation of men"[52] that reflects an economy of human resources. It is these more global reconstructions of our world that should be understood as the gradually emerging result of many small, locally intelligible practices. This result probably could not have been visualized at the outset as their intended end, but its emergence depended upon their contributions to one another.

We should also not regard this emerging pattern as uncontrollable and irresistible. All the tactics and techniques for remaking the world in the image of laboratories, and their strategic results, have encountered resistance. Many of the constraints that must be imposed to extend the controlled environment of a laboratory microworld are ignored or defied by those expected to observe them. (These constraints are also to some extent ignored or defied within the laboratory itself.) The work is done in an easier or sloppier way or is scheduled at the convenience of the workers rather than according to plan. The vigilance necessary to maintain the required isolations and controls is often relaxed. Records are falsified to preserve the appearance of order. There are also organizational difficulties. Materials are not supplied on time, skills are not properly taught, defective equipment is left unrepaired, and pressures are applied to get the job done, even if improperly. The flood of information provided to monitor a complex process may become unmanageable, so that very little of it is utilized. Contradictory goals are sometimes assigned, thus ensuring that not all can be fulfilled. Or the goals imposed from above are subverted from below. The pressures to maintain a more predictable and controllable world are met by people's determination to preserve some unregulated portions of their lives.

There are other sources of resistance and breakdown as well. In entire regions of the world the infrastructure and the social and political arrangements necessary to extend laboratory microworlds have not been provided or maintained. The equipment is not available that would allow things to be measured, purified, isolated, and monitored. Even if it were provided, the skills and practices to use it have not been developed. The result is that attempts to apply scientific knowledge in these contexts often fail. There are other occasions when

52. Foucault 1977: 220.

the outcome of one practice complicates the efforts to maintain the controls and isolations necessary for another (the many forms of environmental pollution are salient examples of this). In still other cases it may be impossible to block off undesired and uncontrolled interactions with the environment and thus to isolate a sufficiently laboratory-like situation. Some processes simply cannot be adequately monitored and controlled; the long-term disposal of radioactive waste or toxic chemicals is probably a good example. Finally, there are sometimes local conflicts between the capillary power relations that must be invoked to extend the laboratory successfully and the interests represented by more traditional sources of power, such as the state or the owners of capital.

We have seen, then, that the attempt to remake and reconceptualize the world on the model of laboratory microworlds is neither a fully coherent project deliberately imposed from above nor an irresistible force that cannot be countered from below by those it affects. It acquires what coherence it has from the way many different local projects and practices coalesce and reinforce one another, producing a situation with an overall meaning and direction that were never anyone's doing in particular. This situation in turn elicits a variety of responses. To a significant extent we all contribute to its continuation, for it shapes the context within which our actions are intelligible even to ourselves. To act otherwise would be madness.[53] But we can also partially recognize its impact upon our lives and resist some of its impositions if we choose. Where such resistance is possible, when it might be called for, and what forms it could take are fundamental political questions for us.

We can now see more clearly why we should describe laboratory practices and their extension as embodying *power* relations. First, there are the overlapping and mutual reinforcement between such practices and the forms of disciplinary power or biopower described by Foucault. I think it is no longer possible, in the light of these arguments, to sustain a political distinction between the exercise of power over human bodies and the development and use of capacities to control and manipulate things. Those who are attracted to Foucault's account of our social practices and institutions as traversed by capillary relations of force or domination will thus have to come to terms with the forms of power/knowledge that invest the natural sciences and our dealings with the physical/biological world. But even those who are

53. For an interesting discussion of this kind of interaction between epistemological and psychiatric categories, see MacIntyre (1980).

dissatisfied with Foucault's analysis cannot really escape dealing with the sciences in terms of power. For the extension of the concept of power to apply in some form to the nontraditionally political institutions of prisons, hospitals, schools, and factories has been much more solidly established than has Foucault's specific treatment of power. And the principal import of my argument in the last two chapters has been that once that extension is accepted, the relationship between these forms of power and what goes on inside scientific laboratories and discourses has become an important issue for political thought.

Understanding scientific practices in terms of their capillary power effects does not depend specifically upon the connection with Foucault's work either. The arguments I have been presenting should now have prepared us to see directly some of the ways scientific practices and their extension beyond the laboratory have introduced new constraints upon our lives, significantly changing what it is possible or intelligible for us to do and imposing new demands upon us. The extension of laboratory practice and understanding has created or transformed our most ordinary, daily rituals (think of washing one's hands to prevent infection, organizing and coordinating activities by the clock,[54] using electric light to detach those activities from the seasonal patterns of daylight, taking and keeping track of medication, and checking the chemical and nutritional composition of foods).

This extension of the laboratory has also led to the institution of massive social and technical systems for utilizing newfound scientific understanding and its capacities for manipulation and control (think of our arrangements for drilling, transporting, refining or recombining, and distributing petroleum products; for creating, distributing, and using alternating-current electricity; for maintaining medical capabilities to monitor, intervene in, and control birth, trauma, and disease; and for providing year-round, nonlocal availability of produce, meat, and dairy products that have been treated and inspected for contamination, spoilage, and quality). These systems demand constant service from us (e.g., supply, maintenance, operation, consumption, accounting) if they are to function readily and reliably; they have also become so integral to our everyday practices that we cannot afford to let them break down.[55] Even as individuals, we can hardly avoid using them, since alternative practices, skills, knowledge, and equipment have often disappeared or been institutionally restricted,

54. Clocks obviously predate modern scientific laboratories. But in addition to contributing new and more exacting measures of time, laboratories have given much greater impetus to the *timing* of what we do, and hence to the use of clocks.

55. Winner 1977, chap. 5.

or else have become inadequate to what we would demand of them.[56] They impose constraints upon us and compel actions from us in ways arguably far more extensive and intensive than the more traditionally recognized forms of political domination.

These constraints have a very different shape from traditionally understood power relations, however. They do not take only the form of a power *over* particular things or persons, possessed and exercised *by* others. They circulate throughout our relations with one another and our dealings with things and pervade the smallest and most ordinary of our doings. They shape the practical configuration within which our actions make sense, both to ourselves and to one another. This shaping occurs most directly through their effects upon the kinds of equipment available to us, the skills and procedures required to use that equipment, the related tasks and equipment that use imposes upon us, and the social roles available to us in performing these tasks. Through this process, they change our understanding of ourselves and our lives.

It is not so much that these power relations compel specific actions from us (although they certainly do that) as that they reconfigure the *style* and interconnectedness of what we do. There are still many possible actions and self-interpretations open to us. But these are drawn from a field of possibilities that increasingly displays characteristic features reminiscent of laboratory microworlds. Both what we do and the things we encounter and employ are more often functionally enclosed and isolated or partitioned. Both are also more carefully measured, counted, timed, and scheduled. Whatever we do, we have become more visible and accountable, with more documentary traces of who we are, where we have been, and what we have done. It is not just that there is more surveillance of our goings-on; we and the things we deal with have been increasingly compelled to *produce* signs of our presence and behavior.

Understanding the world as a collection of resources has introduced an additional set of trade-offs and calculations to the configuration of our activities: questions of efficiency, scarcity, and resource consumption have become involved in more of what we do. Almost anything can be subjected to resource calculations: material resources, certainly, but also energy, time, space, skill, labor ("manpower"), capital, wilderness, daylight, genes, and almost anything else that can be used or stored up for future use. Our activities, organizations, and equipment have also been made more tightly coupled and

56. Winner 1977, chaps. 5–6; Illich 1973, 1981.

more artificially complex (they have more connections to more other actions in more other places). The latter has occurred at the same time that the environment in which they function has been more simplified or closed off. (This isolation includes the social environment as well as the natural one; even as our actions become more complexly interconnected, they are increasingly separated from one another to try to prevent unintended and uncontrolled interactions.) As I noted earlier, this construction of complex organized actions in simplified environments is an effect of the attempt to make causal connections in the world more calculable.

The reflections in the last part of chapter 3 should remind us also that we cannot reconfigure the field of our possible actions in these ways without changing our understanding of who we are and what is at stake in our lives. We understand ourselves in terms of the world we encounter around us, which provides the context within which we work out who we are. The most interesting effect here is the ambiguous sense of ourselves as knowing subjects and known objects, as powerful agents and docile targets of power, as constrained and manipulated but also as the ones for whose sake the manipulation is performed and whose values it supposedly implements. Scientific knowledge is our possession, reflecting the choices and judgments we have made. We exercise the vast manipulative capabilities it provides, and if the results do not fulfill our aims we have only ourselves to blame. But the preceding arguments have also shown that scientific knowledge is embodied in practices that in a sense come to possess us. Along with the local knowledge and capability that science continually provides has come a variety of more global, strategic effects that were not deliberately sought and that we may have only limited power to resist or change. These effects, some of which I have tried to articulate in this chapter, are what I see as politically at issue in our scientific practices. The power relations that interest me are not the power *over* nature that scientific practices provide us, although this aspect has also figured prominently in my account. I have been arguing instead that it is the effects of these practices upon us and our form of life that need to be understood in terms of power and that call for an explicitly political interpretation and criticism. The forms of power/knowledge embodied in the natural sciences help reshape our world as a field of possible action and fundamentally influence what is at stake in our lives.

Epilogue

Toward a Political Philosophy of Science

WE have seen that attempts to use epistemological, metaphysical, or hermeneutical considerations to isolate the natural sciences from any essential political engagement have failed. Modern scientific practices are political in ways that are central to their epistemic success. The construction and theoretical reckoning of causally analyzable and calculable microworlds has contributed both to the vast expansion of our scientific capabilities and to the fundamental recasting of our political situation. It is not just that the application of these capabilities is political. The experimental and theoretical practices of the sciences are themselves forms of power. If we are to acquire an adequate understanding of these developments, we need to construct an explicitly political philosophy of science, aimed at providing the resources for a critical assessment of the political dimension of scientific practices.

This need inevitably leads us to ask what such a political philosophy of science would look like. What issues would it deal with and what problems would it confront? Part of the answer undoubtedly is that many of the current concerns of philosophers of science would remain central to the field. Philosophers must still be concerned with the justification of knowledge claims, the ontological status of the entities and processes described in those claims, and the philosophical problems and opportunities presented by specific developments in the various sciences. Recognizing the practical, social, and political dimensions of these issues places them in a new context, however. Even if this recognition left unchanged some of the specific problems and arguments in these areas of the philosophy of science, it could give their outcome a new significance and urgency.

But new issues would also arise. It may once have seemed plausible that an assessment of the narrowly epistemic rationality of science, or the justification of its claims to knowledge of the natural world, would be sufficient to secure its place above the fray of political debate. We can now see that this view presupposes an untenable isolation of scientific practices from political reality, from the other things we do, and from our understanding of what is at stake in our lives. An understanding of the values and goals internal to scientific practices is indeed crucial to any adequate assessment of their political significance, but this understanding is only the beginning. For what is at issue is whether and how to assess critically the place of these kinds of practices, with these kinds of values and goals, among the many other practices we engage in. We need a perspective from which we can describe and assess what is at stake politically in the way the sciences are practiced and deployed today.

I have not attempted to develop such a perspective within this book. To do so would have required me to provide a much more extensive political context, and a rather different philosophical background, than I have needed to sustain my argument so far. It would also have taken me beyond where I am confident of my own position. The issues here are complex and conflicting, and I do not claim to have a satisfactory response to them. My aim has been primarily to call attention to the political character of our scientific practices and the ways they reach outside the explicitly scientific disciplines and institutions. Such a descriptive project certainly has an implicit political orientation, but I am not yet prepared to work it out or defend it. What I shall do, however, is sketch out several possible approaches to providing such a political perspective. I do so not specifically to recommend any of them or to exclude others, but to give some indication of the directions in which the arguments in this book might point us.

I will consider four approaches. The first is the attempt to incorporate a political interpretation of the sciences within the broad scope of a liberal political theory. Second, I consider the various forms of what I will call "liberationist" criticisms of science. These forms include feminist, Marxist, Third World liberationist, and other criticisms of scientific practices and their political impact, from the perspective of various oppressed groups. The third approach takes the interaction between scientific and political concerns as itself the problem. Writers such as Habermas and Hannah Arendt argue that we have lost any genuine sense of *political* action because political institutions and practices have been usurped by a managerial concern with technical and organizational problems. Finally, we must consider a deeper concern raised in different ways by Heidegger and Foucault. Both see dangers in the forms of power at work in the sciences and technologies, but they are deeply pessimistic about the prospects for a general political assessment of them. This pessimism arises because they believe the available forms of criticism tend to reinforce the very practices and interpretations that are dangerous. They see at best the prospect of a variety of local forms of resistance to power, which do not aim toward a general strategy or even a general justification for resistance. These four approaches are largely incompatible, but the issues raised among them, and presumably by other positions as well, should be central concerns of a political philosophy of science.

Liberal theory is a good place to begin. It is a treacherous undertaking to try to identify the most fundamental strands of the complex liberal political tradition, for almost any such formulation will seem hopelessly inadequate to at least someone within that tradition. Nevertheless, I can point out at least four elements essential to my use of the term. First, the principal issue for a liberal account of power is the legitimacy with which power is exercised (rather than, for example, whether it is exercised efficiently and effectively or oppressively). Second, the principal source of legitimacy must be the lives of individuals, who are taken to be capable of setting their own goals and shaping their own lives and who must be permitted the freedom to exercise these capacities. The legitimate exercise of power either must be freely authorized by the individuals it affects or must be demonstrably in the service of their interests as individuals. Third, no individuals occupy a privileged standpoint with respect to which legitimacy is to be assessed. To this extent, liberal theory treats all individuals as formally equal.[1] Finally, power must be legitimated rationally, usually by derivation from rationally defensible principles. These principles have most commonly been either utilitarian or Kantian/contractarian, but these approaches are probably not essential to a liberal view, so long as it preserves the individual basis of legitimacy.

There are several tasks that a liberal political criticism of the sciences would have to perform. Most fundamental would be the in-

1. Note that Rawls's theory of justice, which assigns a material preference to the interests of the least privileged members of society, does so on the basis of a rational judgment by persons in the "original position," in which their individual standing in society cannot be taken into account. Formal equality is therefore preserved.

terpretation and justification of this critical project itself. Liberal theory has traditionally been concerned with the legitimacy of the exercise of power by the state or by specific agents or organizations. It is not well equipped to deal with forms of power that circulate throughout our social relationships rather than being exercised by specific agents. If liberal theory is to have a response to my analysis of scientific power/knowledge, either the forms of power it identifies must be attributed to individuals or groups whose exercise of it can be legitimated (or not), or else the analysis of the legitimacy of power must be considerably broadened in scope without thereby abandoning fundamental liberal tenets. This enterprise does not seem promising to me.

Suppose, however, that this first task were to be accomplished successfully. Two tasks would still remain. Most basic philosophically would be determining what legitimate basis (if any) there is for regulating or restricting scientific practices or their extensions outside the institutions of science and what interests, values, and principles should govern such regulation. Presumably the basis for any such legitimation would be freeing individuals from any illegitimate constraints placed upon their activities by the performance or extension of such practices. A variety of strategies might be available for adaptation here. There are strategies analogous to Rawls's, in which the assessment of whether power (rather than inequality, as Rawls originally discussed) was legitimate would be based upon its effects upon the most disadvantaged members of the society (in this case, perhaps the most *disempowered* members). An alternative approach would be to consider what individual rights are involved here (possibly conflicting ones) and what claims these rights make upon us. Still a third strategy would be some form of utilitarian cost/benefit analysis (particular scientific practices or their extension would be justifiable only if the costs or constraints they impose are proportionate to the benefit expected from them).

The principal advantage a liberal theorist might claim for this approach can be seen in the form of the remaining task, which is the application of the principles of legitimation to particular scientific practices and their extensions. A liberal political philosophy of science would not touch upon science wholesale. It would allow us to distinguish legitimate from illegitimate practices by the effects of specific practices. It seems quite plausible to say that some practices are dangerous or harmful and cannot legitimately be done, that others are beneficial and should not be restricted without powerful justification, and that some may have elements of both, in which case difficult judgments must be made. Furthermore, the basic compatibility of liberal political goals and the goals and values internal to science will be reassuring to everyone for whom the latter goals are so attractive that the idea of a political criticism of science seems dangerous or threatening. It seems plausible that while some attempts to extend scientific practices and interpretations outside the laboratory might be problematic, most specifically scientific practices would not be. Once one gets through the well-trod issues of military research, experimentation on humans, the use of hazardous materials or procedures, and the perhaps more problematic case of animal experimentation, not much of science seems to be affected (of course, collectively these areas account for a considerable portion of research today). Indeed, the most common result of any liberal criticism, apart from these well-known problematic cases, could easily be the defense of scientific research against any politically motivated interference by others (although Paul Feyerabend attempts to justify some political interference in science on what are basically liberal grounds).[2] If this is what a political philosophy of science amounts to, one might well ask what all the fuss was about.

The assessments of particular scientific practices that a liberal theory would require of us are undoubtedly important and would need to be performed in some way in any case, even if liberal theory provides an ultimately untenable basis for them. But there are two notable objections to regarding such concerns as the principal issues for a political philosophy of science. The first argument is well known; it is often objected that the formal equality accorded all individuals in liberal theory masks real forms of domination and inequality that need to be criticized explicitly from the standpoint of those who are oppressed. This argument leads to the various forms of liberationist criticism, which I will discuss in this section. But it may also be argued that important issues arise with respect to the political significance of scientific knowledge that cannot be effectively analyzed in terms of either legitimacy or oppression. These issues lead us toward the third and fourth approaches outlined above, which I will consider in the final two sections of this epilogue.

Liberationist criticism begins with the analysis of oppression. The initial philosophical task is to identify the social class or group whose oppression is at issue and to characterize the forms their oppression takes, the identity of the oppressors, and the interests or values for

2. Feyerabend 1975.

which the oppression is undertaken. The task specific to a liberationist philosophy of science would then be to examine the place of scientific practices within the framework of oppression and also to consider how and to what extent scientific practice could contribute to a social world freed from these forms of oppression. This latter is important, because the natural sciences have often occupied an ambiguous place within liberationist political criticism. As a form of knowledge and power largely controlled by those identified as oppressors, science has often been linked to oppression. Science has provided ideological support and has been a material force in the hands of oppressors. In some cases, notably the oppression of women and racial minorities, scientific practices and their proclaimed results have themselves been direct sources of oppression.[3] At the same time, science has been used as a critical tool to challenge oppression and its ideological justifications and has been thought to provide a resource for attaining a more just social order. Sandra Harding has aptly described this conflict as raising the question whether the problem lies with bad science or with "science as usual."[4] How these conflicts arise can best be seen in their manifestation within specific liberationist approaches.

Conflicts over the place of science within Marxism have a long history. As is well known, Marx claimed scientific status for his own analysis of the capitalist mode of production and its inevitable demise. Within more traditional Marxist criticism, science as a force of production owned and directed by the owners of capital was thought to enhance the power of capital over the living labor of the proletariat. At the same time, however, the growth of productive forces to which the sciences contributed sharpened the fundamental contradiction between increasingly socialized forms of production and the private appropriation of their product, and hence between the growth of social wealth and the impoverishment of the proletariat. To this extent, the sciences were expected to make a crucial contribution to the occurrence of a crisis in the capitalist mode of production. Furthermore, it is important to Marxist analyses that the augmented productive forces created by capitalism be adaptable to use within a just society. This growth in productive capability was regarded as a necessary precondition for creating a genuinely communist society. Over-

3. This includes both the "scientific" justification of sexism and racism and also the performance of experiments or untested clinical procedures on women and racial minorities. As examples of the latter, consider the Tuskegee syphilis experiments or the use of DES, thalidomide, or oral contraceptives.

4. Harding 1986.

all, then, science played a generally positive role within classical Marxism.[5]

It has also been argued, however—most notably by members of the Frankfurt school, especially Herbert Marcuse, Max Horkheimer, and Theodor Adorno[6]—that science and scientific ideals play an ideological role within advanced capitalist societies and that significant changes in our scientific practices and their political role would be required to overcome their contribution to domination and exploitation. The "value freedom" and the independence from any particular political standpoint that are usually thought essential to the sciences were interpreted as in practice allowing the sciences to serve the interests of the dominant classes in the society. A place was needed for a *critical* theory of society, including within its scope our collective understanding of and dealings with the natural world. These theorists went on to argue that there are internal connections between the "domination of nature," which is central to the understanding and practice of modern science, and the political domination of human beings. Only if we were to develop a more reciprocal and dialogic scientific practice would we be able to respond freely and respectfully to one another.

As I suggested in the Preface, an adequate understanding of scientific practice has often been missing from such political analyses of the sciences. This lack in turn has weakened the analyses of their effects upon us and our social practices and institutions. There are obviously some issues at stake in the different interpretations of science within Marxist theory (and mutatis mutandis, in the differing analyses of science within other liberationist theories) that do not depend specifically upon a detailed understanding of the sciences and of the forms of power/knowledge developed within them. But an inadequate understanding of science is too often a limiting factor within a liberationist philosophy of science.

A good example of this inadequacy has been the frequent failure to take account of the ways power relations invest scientific practices themselves. The focus of liberationist criticism has too often been simply the ideological content of particular scientific theories. Such criticism has been important and valuable in many cases. The attempts to use science to show that patterns of racial or sexual oppression have a biological justification and the surreptitious import of unquestioned political or gender-based metaphors and assumptions

5. For examples of traditional Marxist analyses of science, consider Bernal (1971) and the collection of papers in *Science at the Cross Roads* (1931).

6. Marcuse 1968; Horkheimer and Adorno 1972.

into supposedly apolitical scientific arguments have been justly criticized. So have the attempts to stamp the imprimatur of science upon disciplines whose principal raison d'être is exploitation (e.g., Frederick Taylor's "scientific management"). But many fields of science, including most of the traditionally prestigious fields of mathematics, physics, and chemistry, often seem relatively free of such straightforwardly ideological interests.[7] The focus upon ideology reinforces the idea that there is an essentially apolitical core to the sciences that is immune to criticism. Once we recognize the ways scientific practices themselves express power relations, however, this illusion disappears. A political assessment of the sciences cannot stop at the point where human beings cease to be the immediate object of scientific study, and where ideology ceases to be the principal political concern.

When we turn from Marxist theories to feminist theories, we discover a more complicated ambivalence about the place of the sciences within a liberationist politics. The predominant exclusion of women from science, and the historical identification of scientific thinking and practice with masculinity, have long made science problematic for feminists. The same is true of the frequent neglect of women's concerns, or of knowledge of women, within the sciences and the many ways scientific claims have been used to buttress sexism. But there has been significant disagreement over whether this problem is intrinsic to science as an androcentric practice or whether it represents a failure to maintain genuine scientific objectivity and hence a failure to achieve scientific knowledge. Many of the manifestations of sexism and androcentrism in the sciences have been effectively criticized from the vantage point of scientific ideals. But this criticism has prompted further reflection upon the ideals themselves. Does objectivity require a rigorously gender-neutral perspective? Or is objectivity increased by giving specific critical weight to a feminist standpoint? Or is the very conception of "objectivity" itself an androcentric construct that must be revised fundamentally or abandoned if women are not to be excluded from and oppressed by scientific practice?[8]

This last concern has been given additional impetus by much recent feminist work as well as by contemporary philosophy of science. Psychoanalytically based "object relations" theory has suggested that the concept of "objectivity" as it usually functions in the sciences is deeply rooted in the masculine construction of ego identity within a culture

7. Statistics has been an important exception, as a mathematical field that has had considerable political and ideological importance. Cf. MacKenzie 1981.

8. For a discussion of these issues, see Keller (1985) and Harding (1986).

in which early child rearing is predominantly done by women. This idea has important connections to my analysis of the relation between knowledge and power in the sciences. The feminist appropriation of object relations theory suggests that the connection between what counts as knowledge and the ability to manipulate and control the things known, is culture bound and gender bound. If so, perhaps there does not have to be a link between scientific knowledge and the domination of nature, reflecting a patriarchal and androcentric culture. Women's experience and understanding, if developed, might have the potential to provide a less androcentric, less power oriented, and ultimately more adequate scientific knowledge than the objective, "disinterested," manipulative models of science with which we are familiar. There are problems with this "feminist standpoint" approach to a philosophy of science that have been usefully discussed by Harding and Keller. There is also implicitly a challenge for the advocates of such a feminist approach to science to show concretely what this approach would look like, perhaps even ideally to provide an actual program of research as an alternative for feminist scientists, not just feminist theorists of science.[9]

This last concern brings back the ambivalence lurking within almost all liberationist philosophies of science. Scientific practice as we know it, whatever its political faults, is a powerful force to be reckoned with, and it satisfies some important needs and desires, both intellectual and material. It increasingly attracts people who wish to identify themselves as both feminists and scientists. This situation places an implicit burden upon critics of the scientific enterprise to recognize the importance of those satisfactions. Such recognition would require either an alternative way to provide them or a political criticism that leaves the most basic achievements of the sciences unthreatened. Those who reject this burden and suggest that scientific knowledge

9. Keller 1985; Harding 1986. With respect to the last point, in chapter 6 of her book Harding surveys several attempts to articulate what a specifically feminist science would look like. She rightly points out and objects to the fact that the absence of a fully developed feminist "successor science" is often used to dismiss feminist criticism out of hand. She notes that our current understanding of scientific methodologies was worked out over several centuries, and that it is inappropriate to expect a feminist successor science to resolve comparable problems immediately. Furthermore, she insists that a feminist science must be informed by a more developed criticism of science as it is now practiced; this criticism cannot be abandoned because no full-blown successor project is in sight. Nevertheless, this point should not obscure the fact that the shape of a successor project must be an important concern for feminist criticism and belongs at the center of discussion (it is, obviously, an important theme for Harding's own analysis). The more concrete this discussion, the more effective this dimension of a feminist criticism will be.

itself is deeply tainted with unacceptable political practices and ideals run the risk of marginalization. Evelyn Fox Keller noted of a feminism that turns its back upon objectivity in any form: "By rejecting objectivity as a masculine ideal, it simultaneously lends its voice to an enemy chorus and dooms women to residing outside of the realpolitik modern culture; it exacerbates the very problem it wishes to solve."[10] This potential dilemma lies within any radical feminist (or other liberationist) criticism of the sciences. The problem is how to do justice to the discovery of forms of bias, exclusion, and oppression it identifies within many scientific practices and claims without having to reject wholesale the scientific enterprise and its achievements.

Similar issues must arise from within criticisms of scientific practices and their extensions from the perspective of Third World societies. Western scientific practices have played an important role in the cultural imperialism frequently charged by Third World critics. Furthermore, the universalism usually ascribed to scientific knowledge and practice by its defenders often rings hollow in a Third World context. Scientific research is capital intensive. It also benefits greatly from the close interaction of capable practitioners, which encourages them to congregate in Western institutions. Third World countries commonly do not have the resources to support Western-style scientific research (these costs are also usually much higher in such nations than elsewhere, because of increased expenditures for importing scientific instruments and supplies and for travel to meet peers) or the contacts and critical mass of researchers even to keep in close touch with developments at the forefront of science. As a result, the attempt to train scientists in the Third World more frequently brings the loss of intellectual talent to the West than the achievement of excellence in science at home. Yet Western scientific institutions are often thought to be uninterested in problems of specific interest to Third World nations or at least unwilling to devote resources to finding solutions applicable to them. To the extent that this criticism is true, the sciences continually extract some of the most talented members from Third World societies without providing anything like adequate recompense. This is, of course, a pattern in First World/Third World relations that has not been limited to the exploitation of human resources.

Other problems also arise with the attempt to utilize scientific knowledge within a Third World context. The example of the green revolution was discussed in the previous chapter. But the spread of

10. Keller 1982: 593.

acquired immune deficiency syndrome (AIDS) in Central Africa illustrates the problems even more clearly. Transfusions from contaminated blood supplies are a major means of transmitting the disease, which now seems to be reaching epidemic proportions. The use of blood typing and transfusions represented an admirable attempt to extend some of the basic capabilities of Western medical science to the rest of the world. Nevertheless, in countries that cannot afford enough hypodermic needles to meet the demand, and in which a single test for antibodies to the AIDS virus (the basic technique for avoiding infected blood supplies) costs several times the national per capita expenditure on all health care, the value of such practices may be illusory.[11] Similar difficulties were evident in the marketing of powdered infant formula outside the West and in the production and use of drugs and toxic chemicals. The problems go back to the controls needed to allow scientific practices to be extended outside the laboratory environment and the economic and social costs of providing them in societies that have not already been reconstructed to make this possible.

Science has often been touted as a key to the advancement of "underdeveloped" countries, and its attractiveness in this role is real and important. But it may also make false promises and even undermine the attempts of Third World societies to confront their problems in ways accessible to their resources and appropriate to their traditions. There may even be a deeper problem when we consider the role of the sciences in maintaining dependence upon the West and the ways the limited importation of scientific knowledge and practices can sometimes wreak havoc in a society. Suppose it were *not* possible either to extend the scientific reconstruction of the world everywhere or to insulate less "developed" societies from the fallout of scientific/technical societies elsewhere (disproportionate resource consumption, environmental degradation, economic dependence, etc., as well as the kinds of problems just alluded to). We would then have to confront seriously with respect to scientific practices and their widespread extension outside the laboratory a moral challenge proposed by Ivan Illich: "The principal source of injustice in our epoch is political approval for the existence of tools that by their very nature restrict to a very few the liberty to use them in an autonomous way."[12] If this criticism were to prove true of the principal equipment and

11. Altman 1985.
12. Illich 1973: 46.

practices of the sciences, a Third World liberationist critique of these practices would cut deep indeed.

A very different sort of political criticism of the sciences is raised by those who object to the degradation of political action by the encroachment of purely technical or managerial concerns into the sphere of politics. I discussed in chapter 6 Habermas's attempt to distinguish the cognitive interests satisfied by science and technology, on the one hand, and by political action and communication, on the other. His aim in drawing this distinction was to protect the sciences from political criticism *within the sphere of their appropriate use* while challenging the importation of technical thought and practice into what should have remained the domain of political and communicative action.[13] A similar program, from a somewhat different philosophical orientation, was put forward by Hannah Arendt in *The Human Condition.* She argued that the transformation of natural into universal science, and of action into labor, has reduced and impoverished the possibilities for action, which alone can reveal us in our human distinctness. Arendt is rather less inclined to defend the modern form of the sciences than is Habermas, for she seems to feel that the modern turn toward a universal science cannot help but encroach upon the political domain.[14] Nevertheless, both are concerned to restore the practical significance of a meaningful realm of human affairs and relationships, distinct from our pragmatic dealings with things, and both implicate the growth of the natural sciences in the eclipse of the practical realm within the modern world.

I argued in chapter 6 against attempts like Habermas's to distinguish an interpretive field of human action and meaning from the natural world, which contains events but not actions. These arguments do not by themselves undermine this approach to a political philosophy of science, however. Criticisms by Arendt, Habermas, and others of the form and content of our political life, of its alleged decline into technical or bureaucratic management, may still represent a powerful objection to the way modern societies have developed. What my arguments do imply is that the practices of the natural sciences cannot be partitioned off from the criticism. If there has been a loss of meaning or of vital significance in our political life, then its reinvigoration must involve criticism of our scientific practices and

13. Habermas 1968b, 1970.
14. Arendt 1958.

their power relations as well. Unless these arguments can be shown to be inadequate, the challenge to such a political philosophy of science is straightforward: along with its critique of political practice, it must articulate a vision of a renewed political understanding and practice that would also include a transformation of the sciences consistent with the renewal. Like the more radical feminists, philosophers in the tradition of Habermas and Arendt may have to make some difficult choices about what aspects of our scientific practices can and should be preserved within their view of a more adequate politics.

The final approach I shall consider, which is exemplified in different ways in the work of Heidegger and Foucault, puts in question the very notion of political criticism. We are accustomed to understanding such criticism as based on a normative standpoint with respect to which the criticism is made. Thus a liberal approach takes as basic the worth of free, autonomous self-determination by individuals and criticizes social institutions and practices according to whether they enhance or constrain such autonomy; liberationists begin with the interests of oppressed groups and try to determine what contributes to oppression and what resists it in order to support the resistance; and Habermas and Arendt both insist upon the importance of communicative action for a fully human social order and aim to restore and enrich the possibilities for such action. Both Heidegger and Foucault see serious dangers in the configuration of practices within which we are situated today, but they resist formulating these dangers into a normative basis for political criticism. The reason is that, in somewhat different ways, they believe the understanding of political engagement as normative contributes to the dangers to which they aim to respond. If we are to negotiate a way through these dangers, they insist, we must engage them in ways that do not take the form of a normative political criticism.

For Heidegger the danger is nihilism. If the nihilistic propensity in our practices were to be fulfilled, nothing would seriously matter to us anymore, and thus nothing would be at stake in what we do. We encountered this notion in chapter 6, where we saw Dreyfus try to use it to establish a fundamental difference between human and natural sciences. Heidegger, however, sees the danger of nihilism lurking not only in the ways we have tried to objectify human beings, but also in our understanding of the things around us as calculable, manipulable resources. In either case, thinking is interpreted as calculating representation and action as willful bringing about of some end.

Why is this position nihilistic? When we understand the world as composed of resources that can be used in interchangeable and more or less efficient ways, we discover choices between alternative ways of using them. These choices can be made reasonably only with respect to what our resources are to be used *for.* But this itself has to be determined, and the only apparent way to do so is through further calculative theoretical reflection. What is at issue for us must therefore be resolved into materials for such calculation, that is, into alternative *values* that can be clarified, criticized, and chosen. This is the significance of the concept of "value": it transforms the configuration of practices within which thought and action are intelligible to us into something we can reckon with and willfully implement. When this happens, however, the values chosen do not govern our choice, for they are what is chosen. That for the sake of which our choice of values is made always withdraws from calculative awareness. This complication calls for further reckoning, in a futile attempt to disclose our "ultimate" values. Heidegger depicts an endless expansion of calculative thought and practice, as we subject the world to increasingly precise manipulation and control and relentlessly try to get a fix upon who we are and why we are doing this. What turns out to be at issue here is the increase of power itself. Our practices are focused by a will to power, a continual striving for increased control and more precise determination of ourselves and the world, that is *never* subordinated to any other concern. Heidegger noted in his lectures on Nietzsche: "Power can maintain itself in itself, that is, in its essence, only if it overtakes and overcomes the power level it has already attained. . . . 'Will to power' does not mean simply the 'romantic' yearning and quest for power by those who have no power; rather, 'will to power' means the accruing of power by power for its own overpowering."[15] It is this that governs our clarification and choice of values. The attempt to acquire mastery over ourselves and what is at issue in our practices is the epitome of this will to power: "The new valuation . . . supplants all earlier values with power, the uppermost value, but first and foremost because power and *only power* posits values, validates them, and makes decisions about the possible justifications of a valuation. . . . To the extent that it is truly power, alone determining all beings, power does not recognize the worth or value of anything outside of itself."[16]

The problem is not that everything does become a calculable resource available to be used and that nothing is really at issue in our practices. The central insight of a practical hermeneutics—that what

15. Heidegger 1982: 7.
16. Heidegger 1982: 7.

is at stake in our practices is not up for choice—has not been challenged. We have not *chosen* to focus our practices in the will to power; we find all our possibilities situated within it. What makes this situation nihilistic is that there is no stable content to what is at issue for us. Will to power is the constant *annihilation* of any stable field of meaning that might provide coherence to our lives. As David Kolb has noted, Heidegger's concern is thus reminiscent of Hegel's critique of modernity as an age of absolutely free subjectivity that affirms itself only by abolishing any concrete content to its freedom, except that Heidegger does not locate this empty willing in the individual subject.[17]

This is not the place to go into the complexities of Heidegger's response to this situation. What is important for our purposes is that if he is right, any general attempt to establish values with respect to which our situation could be criticized only makes things worse. Whatever our situation calls for must be enacted in local responses to whatever concrete possibilities do emerge for us as meaningful. We have no basis for projecting what will result from this or, indeed, whether anything other than the increased domination of the will to power will result. But such local responsiveness to what comes our way is the only hope Heidegger can offer.

Foucault arrives in a somewhat different way at a similar rejection of global normative justifications for political engagement. We have already seen important elements of his account of the modern forms of power. Foucault has described the use of surveillance and coercion to create docile, normalized discursive subjects. He objects to the oppressive *effects* of these power relations, and his descriptions aim to show clearly their oppressiveness. His concern is not to assess the legitimacy or illegitimacy of these deployments of power, but to galvanize effective *resistance* against the particular forms of power he finds oppressive.

Foucault finds liberationist forms of political criticism unacceptable because they attempt to situate themselves outside the power relations he finds pervasive and inescapable. Liberationist critics attempt to identify the real interests of the oppressed and to ground their analyses in an understanding of the global structure of oppression. Their analyses are supposed to reflect a knowledge freed from the distortions of power, in contrast to the ideological and oppressive claims to knowledge they criticize. Foucault not only thinks this supposition is mistaken, he regards it as a contribution to the very forms of power he wishes to combat.

To see this point, it is important to remember that Foucault does

17. Kolb 1981.

not think the modern forms of power have entirely supplanted the earlier forms of juridical power. The latter not only continue to operate, they also mask the disciplinary power that operates at a micro-level of social relations. Thus the liberationist claims to freedom from the distorting effects of power simply allow the old forms of oppression to continue in a new guise. The issue was effectively highlighted in a discussion between Foucault and several Maoist activists concerning the notion of "popular justice." The Maoists objected to the administration of justice by the bourgeois state but contrasted this to the judicial procedures of a "revolutionary state apparatus" exemplified in the Chinese revolution, which supposedly implement a genuine "people's justice." Foucault argued that a people's court was still a court, with all its basic power relations left intact. The imprimatur of revolutionary justice only masks this maintenance of the traditional forms of oppression. Consequently, he concludes:

> The revolution can only take place via the radical elimination of the judicial apparatus, and anything which could reintroduce the penal apparatus, anything which could reintroduce its ideology and enable this ideology to surreptitiously creep back into popular practices, must be banished. This is why the court, an exemplary form of this judicial system, seems to me to be a possible location for the reintroduction of the ideology of the penal system into popular practice. This is why I think that one should not make use of such a model.[18]

Liberationist criticisms, he thinks, overlook as sources of oppression the ways power is *administered* and consequently entrench them all the more effectively. This is presumably the case whether we are talking about the courts and prisons, the schools, the hospitals, or the laboratories.

Foucault traces the fault to the *global* character of any normative criticism, of which he takes Marxist or psychoanalytic theories to be salient examples. Foucault sees such global criticism as an attempt to reinstitute the traditional legitimation of power through claims to knowledge. The normative/theoretical framework that supposedly underwrites political criticism itself becomes immune to opposition and hence mimics the forms of power/knowledge against which Foucault aligns his arguments.

> Which theoretical-political avant garde do you want to enthrone in order to isolate it from all the discontinuous forms of knowledge that circulate about it? When I see you straining to establish the scientificity

18. Foucault 1980: 16.

> of Marxism . . . for me you are doing something altogether different, you are investing Marxist discourses and those who uphold them with the effects of a power which the West since Medieval times has attributed to science and has reserved for those engaged in scientific discourse.[19]

Foucault urges as an antidote to a normative, theoretical political criticism a resurrection of the "local, subjugated knowledges" that reflect an intimate acquaintance with the force of disciplinary power. Such knowledges oppose power directly in its contemporary forms rather than aiming to appropriate that power in the service of a new truth. Thus he says of his own recent inquiries:

> Let us give the term *genealogy* to the union of erudite knowledge and local memories which allows us to establish a historical knowledge of struggles and to make use of this knowledge tactically today. . . . What [genealogy] really does is to entertain the claims to attention of local, discontinuous, disqualified, illegitimate knowledges against the claims of a unitary body of theory which would filter, hierarchise and order them in the name of some true knowledge and some arbitrary idea of what constitutes a science and its objects.[20]

Only such local forms of resistance can hope to avoid being co-opted by the predominant forms of power, Foucault believes. David Hiley describes this aptly as a "political engagement without liberal hope or comfort."[21] Foucault offers us no guarantee that we are engaging in the "right" struggles, in the service of the true interests of humanity; nor does he provide the satisfaction of a communal struggle, of the Great March of humanity working together to liberate itself.[22] He only suggests that we may thereby avoid surreptitiously contributing to the tightening of the disciplinary constraints that he sees administering our lives at every turn. Foucault, then, would call for a political engagement with the sciences that dispenses with a political philosophy of science to justify it. We would, however, need some further argument assessing what if anything is dangerous or oppressive in the power relations issuing from the sciences before we could specify what concrete forms such political engagement might take.

These considerations may seem to have taken us a long way from my central concern to understand what happens in scientific laborato-

19. Foucault 1980: 85.
20. Foucault 1980: 83.
21. Hiley 1984.
22. Kundera 1984.

ries and how the sciences affect us in our dealings with the world and with one another. If we reflect upon the arguments I have tried to develop in this book, however, perhaps it is not such a long way. I have argued that we cannot isolate the scientific enterprise from the myriad other practices through which we define ourselves as human beings. The sciences have shaped the world we live in and the possibilities in terms of which we might come to understand ourselves. There is nothing politically neutral about these effects, though we may well appreciate them primarily as contributions to our well-being rather than as a danger or a threat. But how we come to terms with them must ultimately merge with our most basic understanding and criticism of what is at stake in our lives. It remains to be seen whether this will be couched in terms of individual possibilities for fulfillment, social struggles for liberation, the rejuvenation of genuine prospects for human action and interaction, or some other terms perhaps yet unfamiliar to us. These, however, are the kinds of issues with which a philosophy of science must ultimately concern itself if we are to take seriously the importance of the sciences in the modern world.

References

Ackermann, Robert. 1985. *Data, Instruments, and Theory: A Dialectical Approach to the Philosophy of Science.* Princeton: Princeton University Press.

Allen, Garland. 1975. *Life Science in the Twentieth Century.* New York: John Wiley.

Altman, Laurence K. 1985. *New York Times,* 24 November, sec. 1, p. 34.

Arendt, Hannah. 1958. *The Human Condition.* Chicago: University of Chicago Press.

Bernal, John D. 1971. *Science in History.* 4 vols. Cambridge: MIT Press.

Bernstein, Richard. 1983. *Beyond Objectivism and Relativism.* Philadelphia: University of Pennsylvania Press.

Bleier, Ruth, ed. 1986. *Feminist Approaches to Science.* Oxford: Pergamon Press.

Boyd, Richard. 1973. Realism, Underdetermination, and a Causal Theory of Evidence. *Nous* 7.

———. 1980. Scientific Realism and Naturalistic Epistemology. In *PSA 1980,* vol. 2, ed. Peter Asquith and Ronald Giere. East Lansing, Mich.: Philosophy of Science Association, 613–62.

———. 1984. The Current Status of Scientific Realism. In Leplin 1984: 41–82.

Brackbill, Yvonne, June Rice, and Diony Young. 1984. *Birth Trap.* Saint Louis, Mo.: C. V. Mosby.

Brandom, Robert. 1983. Heidegger's Categories. *Monist* 66 (July): 387–409.

Cartwright, Nancy. 1983. *How the Laws of Physics Lie.* Oxford: Oxford University Press.

Cohen, Jack, and Franklin Portugal. 1977. *A Century of DNA.* Cambridge: MIT Press.

Collins, Harry. 1982. The Replication of Experiments in Physics. In *Science in Context,* ed. Barry Barnes and David Edge. Cambridge: MIT Press.

Commoner, Barry. 1976. *The Poverty of Power.* New York: Alfred Knopf.

Compton, John. 1983. Natural Science and Being-in-the-World. Paper presented to the Pacific Division, American Philosophical Association, March 1983.
———. N.d. Some Contributions of Phenomenology to the Philosophy of Natural Science. Forthcoming.
Condorcet, Marquis de. 1955. *Sketch Towards a Historical Picture of the Progress of the Human Mind.* London: Weidenfeld and Nicolson.
Davidson, Donald. 1984. *Inquiries into Truth and Interpretation.* Oxford: Oxford University Press.
Dilthey, Wilhelm. 1956, 1957. *Gesammelte Schriften.* Vols. 5 and 7. Stuttgart: Teubrier.
Douglas, Jack. 1967. *The Social Meanings of Suicide.* Princeton: Princeton University Press.
Dreyfus, Hubert. 1979. *What Computers Can't Do.* 2d ed. New York: Harper and Row.
———. 1980. Holism and Hermeneutics. *Review of Metaphysics* 34: 3–23.
———. 1981. Knowledge and Human Values: A Genealogy of Nihilism. *Teachers College Record* 82: 507–20.
———. 1984. Why Current Studies of Human Capacities Can Never Be Made Scientific. *Berkeley Cognitive Science Report* 11: 1–17.
Dreyfus, Hubert, and Stuart Dreyfus. 1986. *Mind over Machine.* New York: Free Press.
Dreyfus, Hubert, and Paul Rabinow. 1983. *Michel Foucault: Beyond Structuralism and Hermeneutics.* 2d ed. Chicago: University of Chicago Press.
Farrington, Benjamin. 1951. Temporis Partus Masculus: An Untranslated Writing of Francis Bacon. *Centaurus* 1: 193–205.
Feyerabend, Paul. 1975. *Against Method.* London: New Left Books.
Fine, Arthur. 1986. *The Shaky Game: Einstein, Realism, and the Quantum Theory.* Chicago: University of Chicago Press.
Fleck, Ludwik. 1979. *Genesis and Development of a Scientific Fact.* Chicago: University of Chicago Press.
Follesdal, Dagfinn. 1979. Hermeneutics and the Hypothetico-Deductive Method. *Dialectica* 33: 319–36.
Foucault, Michel. 1973. *Madness and Civilization: A History of Insanity in the Age of Reason.* New York: Random House.
———. 1975. *Birth of the Clinic: An Archaeology of Medical Perception.* New York: Random House.
———. 1977. *Discipline and Punish.* Trans. Alan Sheridan. New York: Random House.
———. 1978. *The History of Sexuality.* Vol. 1. New York: Random House.
———. 1980. *Power/Knowledge.* New York: Pantheon.
———. 1983. The Subject and Power. In Dreyfus and Rabinow 1983: 208–26.
Frankel, Francine. 1971. *India's Green Revolution: Economic Gains and Political Costs.* Princeton: Princeton University Press.
Geertz, Clifford. 1979. Deep Play: Notes on the Balinese Cockfight. In Rabinow and Sullivan 1979: 181–223.
Geuss, Raymond. 1981. *The Idea of a Critical Theory.* Cambridge: Cambridge University Press.

Glymour, Clark. 1982. Conceptual Scheming, or Confessions of a Metaphysical Realist. *Synthèse* 51: 169–80.
Goodfield, June. 1981. *An Imagined World.* New York: Harper and Row.
Goodman, Nelson. 1978. *Ways of Worldmaking.* Indianapolis: Bobbs-Merrill.
Guillemin, Roger. 1963. Sur la nature des substances hypothalamiques qui controlent la sécrétion des hormones antehypophysaires. *Journal de Physiologie* 55: 7–44.
Gutting, Gary. 1979. Continental Philosophy of Science. In *Current Research in Philosophy of Science,* ed. P. D. Asquith and H. E. Kyburg. East Lansing, Mich.: Philosophy of Science Association.
———. 1980. *Paradigms and Revolutions.* Notre Dame, Ind.: University of Notre Dame Press.
Habermas, Jürgen. 1968a. *Erkenntnis und Interesse.* Frankfurt am Main: Suhrkamp.
———. 1968b. *Technik und Wissenschaft als "Ideologie."* Frankfurt am Main: Suhrkamp.
———. 1970. *Toward a Rational Society.* Trans. Jeremy Shapiro. Boston: Beacon Press. Translation of Habermas 1968b.
———. 1971. *Knowledge and Human Interests.* Trans. Jeremy Shapiro. Boston: Beacon Press. Translation of Habermas 1968a.
Hacking, Ian. 1982. Language, Truth and Reason. In *Rationality and Relativism,* ed. M. Hollis and S. Lukes, 48–66. Cambridge: MIT Press.
———. 1983. *Representing and Intervening.* Cambridge: Cambridge University Press.
———. 1984. Five Parables. In *Philosophy in history,* ed. Richard Rorty, Jerome Schneewind, and Quentin Skinner, 103–24. Cambridge: Cambridge University Press.
Hanson, N. R. 1958. *Patterns of Discovery.* Cambridge: Cambridge University Press.
Harding, Sandra. 1986. *The Science Question in Feminism.* Ithaca: Cornell University Press.
Harding, Sandra, and Merrill Hintikka, eds. 1983. *Discovering Reality.* Dordrecht: D. Reidel.
Heelan, Patrick. 1983. *Space-Perception and the Philosophy of Science.* Berkeley: University of California Press.
Heidegger, Martin. 1952. *Holzwege.* Frankfurt: Vittorio Klostermann.
———. 1954. *Vorträge und Aufsätze.* Pfullingen: Gunther Neske.
———. 1957. *Sein und Zeit.* Tubingen: Neomarius.
———. 1959. *Unterwegs zur Sprache.* Pfullingen: Gunther Neske.
———. 1962. *Being and Time.* Trans. John MacQuarrie and Edward Robinson. New York: Harper and Row. Translation of Heidegger 1957.
———. 1971. *On the Way to Language.* Trans. Peter Hertz. New York: Harper and Row. Translation of Heidegger 1959.
———. 1977. *The Question Concerning Technology and Other Essays.* Trans. William Lovitt. New York: Harper and Row. Translated from Heidegger 1952, 1954.
———. 1982. *Nietzsche, Volume IV: Nihilism.* Trans. Frank A. Capuzzi. New York: Harper and Row.

Hesse, Mary. 1974. *The Structure of Scientific Inference.* Berkeley: University of California Press.
———. 1980. *Revolutions and Reconstructions in the Philosophy of Science.* Bloomington: University of Indiana Press.
Hiley, David. 1984. Foucault and the Analysis of Power: Political Engagement without Liberal Hope or Comfort. *Praxis International* 4, no. 2: 192–207.
Hollis, Martin, and Steven Lukes. 1982. *Rationality and Relativism.* Cambridge: MIT Press.
Horkheimer, Max, and Theodor Adorno. 1972. *Dialectic of Enlightenment.* Trans. John Cumming. New York: Seabury Press.
Horwich, Paul. 1982. Three Forms of Realism. *Synthèse* 51: 181–201.
Hunter, J. S. 1980. The National System of Scientific Measurement. *Science* 210: 869–75.
Husserl, Edmund. 1954. *Die Krisis der Europäischen Wissenschaften und die transcendentale Phänomenologie.* The Hague: Martinus Nijhoff.
———. 1970. *The Crisis of European Sciences and Transcendental Phenomenology.* Trans. David Carr. Evanston, Ill.: Northwestern University Press. Translation of Husserl 1954.
Ihde, Don. 1979. *Technics and Praxis.* Dordrecht: D. Reidel.
Illich, Ivan. 1973. *Tools for Conviviality.* New York: Harper and Row.
———. 1981. *Shadow Work.* Salem, N.H.: Marion Boyars.
Kant, Immanuel. 1965. *Critique of Pure Reason.* Trans. Norman Kemp-Smith. New York: St. Martins.
Keller, Evelyn Fox. 1982. Feminism and Science. *Signs* 7: 589–602.
———. 1985. *Reflections on Gender and Science.* New Haven: Yale University Press.
Kierkegaard, Sören. 1954. *Fear and Trembling* and *The Sickness unto Death.* Trans. Walter Lowrie. Princeton: Princeton University Press.
Knorr-Cetina, Karin. 1981. *The Manufacture of Knowledge.* Oxford: Pergamon Press.
Kolb, David. 1981. Hegel and Heidegger as Critics. *Monist* 64 (October): 481–99.
Koyre, Alexandre. 1957. *From the Closed World to the Infinite Universe.* Baltimore: Johns Hopkins University Press.
Kuhn, Thomas. 1957. *The Copernican Revolution.* Cambridge: Harvard University Press.
———. 1970a. *The Structure of Scientific Revolutions.* 2d ed. Chicago: University of Chicago Press.
———. 1970b. Reflections on My Critics. In *Criticism and the Growth of Knowledge,* ed. Imre Lakatos and Alan Musgrave, 231–78. Cambridge: Cambridge University Press.
———. 1977. *The Essential Tension.* Chicago: University of Chicago Press.
Kundera, Milan. 1984. *The Unbearable Lightness of Being.* Trans. Michael H. Heim. New York: Harper and Row.
Lakatos, Imre. 1978. *The Methodology of Scientific Research Programmes.* Cambridge: Cambridge University Press.

Lappe, Francine, and Joseph Collins. 1977. *Food First: Beyond the Myth of Scarcity.* Boston: Houghton Mifflin.
Latour, Bruno. 1983. Give Me a Laboratory and I Will Raise the World. In *Science Observed,* ed. K. Knorr-Cetina and M. Mulkay, 141–70. London: Sage.
Latour, Bruno, and Steve Woolgar. 1979. *Laboratory Life.* London: Sage.
Laudan, Larry. 1977. *Progress and Its Problems.* Berkeley: University of California Press.
———. 1981. A Confutation of Convergent Realism. *Philosophy of Science* 48: 19–49; reprinted in Leplin 1984: 218–49.
Leiss, William. 1974. *The Domination of Nature.* Boston: Beacon Press.
Leplin, Jarrett, ed. 1984. *Scientific Realism.* Berkeley: University of California Press.
MacIntyre, Alasdair. 1980. Epistemological Crises, Dramatic Narrative and the Philosophy of Science. In Gutting 1980: 54–74.
MacKenzie, Donald. 1981. *Statistics in Britain, 1865–1930: The Social Construction of Scientific Knowledge.* Edinburgh: Edinburgh University Press.
McMullin, Ernan. 1984. The Case for Scientific Realism. In Leplin 1984: 8–40.
Macomber, William. 1968. *The Anatomy of Disillusion.* Evanston, Ill.: Northwestern University Press.
Marcuse, Herbert. 1968. Industrialization and Capitalism in the Work of Max Weber. In *Negations,* 201–26. Boston: Beacon Press.
Merchant, Carolyn. 1980. *The Death of Nature.* San Francisco: Harper and Row.
Minsky, Marvin, and Seymour Papert. 1973. MIT Artificial Intelligence Laboratory Memo no. 299, September.
Musgrave, Alan. 1980. Kuhn's Second Thoughts. In Gutting 1980: 39–53.
Nanney, David L. 1957. The Role of the Cytoplasm in Heredity. In *The Chemical Basis of Heredity,* ed. W. D. McElroy and H. D. Glass. Baltimore: Johns Hopkins University Press.
Neiswanger, Katherine. 1985. Southern Blot Analysis in the Hominoidea of the Insulin Gene and the DNA Region of Tandem Repeats 5′ to the Insulin Gene in Humans. Ph.D. diss., University of California at Los Angeles.
Newton-Smith, W. H. 1981. *The Rationality of Science.* London: Routledge and Kegan Paul.
Niiniluoto, Ilkka. 1977. On the Truth-Likeness of Generalizations. In *Basic Problems in Methodology and Linguistics,* ed. R. E. Butts and J. Hintikka, 121–47. Dordrecht: D. Reidel.
———. 1980. Scientific Progress. *Synthèse* 45: 427–62.
Okrent, Mark. 1984. Hermeneutics, Transcendental Philosophy and Social Science. *Inquiry* 27: 23–49.
———. n.d. *Understanding and Being: Pragmatism in Heidegger's Critique of Metaphysics.* Ithaca: Cornell University Press.
Perrow, Charles. 1984. *Normal Accidents.* New York: Basic Books.
Pickering, Andrew. 1984. *Constructing Quarks.* Chicago: University of Chicago Press.

Pinch, Trevor. 1980. Theoreticians and the Production of Experimental Anomaly: The Case of Solar Neutrinos. In *The Social Process of Scientific Investigation,* ed. Karin Knorr-Cetina, Roger Krohn, and Robert Whitley, 77–106. Dordrecht: D. Reidel.
Polanyi, Michael. 1958. *Personal Knowledge.* Chicago: University of Chicago Press.
Popper, Karl. 1953. *The Logic of Scientific Discovery.* New York: Harper and Row.
Putnam, Hilary. 1978. *Meaning and the Moral Sciences.* London: Routledge and Kegan Paul.
———. 1981. *Reason, Truth and History.* Cambridge: Cambridge University Press.
———. 1982. Why There Isn't a Ready-Made World. *Synthèse* 51: 141–67.
Quine, Willard V. O. 1953. *From a Logical Point of View.* Cambridge: Harvard University Press.
———. 1969. Reply to Chomsky. In *Words and Objections,* ed. Donald Davidson and Jaako Hintikka. Dordrecht: D. Reidel.
———. 1970. *Philosophy of Logic.* Englewood Cliffs, N.J.: Prentice-Hall.
———. 1975. *Ontological Relativity and Other Essays.* New York: Columbia University Press.
———. 1976. *The Ways of Paradox and Other Essays.* Cambridge: Harvard University Press.
Quine, Willard V. O., and Ullian, J. S. 1970. *The Web of Belief.* New York: Random House.
Rabinow, Paul, and William Sullivan. 1979. *Interpretive Social Science: A Reader.* Berkeley: University of California Press.
Ravetz, Jerome. 1971. *Scientific Knowledge and Its Social Problems.* Oxford: Oxford University Press.
Rorty, Richard. 1979. *Philosophy and the Mirror of Nature.* Princeton: Princeton University Press.
———. 1982. *Consequences of Pragmatism.* Minneapolis: University of Minnesota Press.
Rouse, Joseph. 1981. Kuhn, Heidegger and Scientific Realism. *Man and World* 14: 269–90.
Scheffler, Israel. 1967. *Science and Subjectivity.* Indianapolis: Bobbs-Merrill.
———. 1972. Vision and Revolution. *Philosophy of Science* 39: 366–74.
Science at the Cross Roads. 1931. London: Kniga.
Scott, James C. 1976. *The Moral Economy of the Peasant.* New Haven: Yale University Press.
Shapere, Dudley. 1964. The Structure of Scientific Revolutions. *Philosophical Review* 73: 383–94.
———. 1971. The Paradigm Concept. *Science* 172: 706–9.
———. 1982. The Concept of Observation in Philosophy and Science. *Philosophy of Science* 49: 485–525.
Shimony, Abner. 1976. Comments on Two Epistemological Theses of Thomas Kuhn. In *Essays in memory of Imre Lakatos,* ed. Robert Cohen et al., 569–88. Dordrecht: D. Reidel.

Sokolowski, Robert. 1978. *Presence and Absence.* Bloomington: Indiana University Press.
Suppe, Frederick. 1977. *The Structure of Scientific Theories.* Urbana: University of Illinois Press.
Taylor, Charles. 1964. *The Explanation of Behavior.* London: Routledge and Kegan Paul.
———. 1979. Interpretation and the Sciences of Man. In Rabinow and Sullivan 1979: 25–71.
———. 1980. Understanding in Human Science. *Review of Metaphysics* 34: 25–38.
———. 1982. Rationality. In Hollis and Lukes 1982: 87–105.
Todes, Samuel. 1966. A Comparative Phenomenology of Perception and Imagination, part I. *Journal of Existentialism* 6: 253–68.
———. 1969. Sensuous Abstraction and the Abstract Sense of Reality. In *New essays in phenomenology,* ed. J. M. Edie. Chicago: Quadrangle Books.
———. 1975. Shadows in Knowledge. In *Dialogues in Phenomenology,* ed. Don Ihde and Richard Zaner, 94–113. The Hague: Martinus Nijhoff.
Toulmin, Stephen. 1961. *Foresight and Understanding.* New York: Harper and Row.
———. 1972. *Human Understanding.* Vol. 1. Princeton: Princeton University Press.
Van Fraasen, Bas. 1980. *The Scientific Image.* Oxford: Oxford University Press.
Watson, James. 1968. *The Double Helix.* New York: Atheneum.
Will, Frederick. 1981. Reason, Social Practice, and Scientific Realism. *Philosophy of Science* 48: 1–18.
Williams, Michael. 1984. Do We Need a Theory of Truth? Paper presented at the Philadelphia Consortium of Philosophy Departments Conference "Realism, Truth, and the Physical Sciences: Realism and Anti-Realism," Bryn Mawr, Pa., 20 October.
Wilson, Bryan. 1970. *Rationality.* Oxford: Basil Blackwell.
Winner, Langdon. 1977. *Autonomous Technology.* Cambridge: MIT Press.
Winograd, Terry. 1973. A Procedural Model of Language Understanding. In *Computer Models of Thought and Language,* ed. Roger Schank and Kenneth Colby. San Francisco: W. H. Freeman.
Wittgenstein, Ludwig. 1953. *Philosophical Investigations.* Oxford: Basil Blackwell.
Wright, Georg Henrik von. 1974. *Causality and Determinism.* New York: Columbia University Press.

Index